Swarm Intelligence and Evolutionary Algorithms for Real World Applications

Swarm Intelligence and Evolutionary Algorithms for Real World Applications

Guest Editor

Mohammad Majid al-Rifaie

MDPI

Basel • Beijing • Wuhan • Barcelona • Belgrade • Novi Sad • Cluj • Manchester

Guest Editor
Mohammad Majid al-Rifaie
School of Computing and
Mathematical Sciences
University of Greenwich
London
UK

Editorial Office
MDPI AG
Grosspeteranlage 5
4052 Basel, Switzerland

This is a reprint of the Special Issue, published open access by the journal *Algorithms* (ISSN 1999-4893), freely accessible at: https://www.mdpi.com/journal/algorithms/special_issues/592XRDTX6X.

For citation purposes, cite each article independently as indicated on the article page online and as indicated below:

Lastname, A.A.; Lastname, B.B. Article Title. *Journal Name* **Year**, *Volume Number*, Page Range.

ISBN 978-3-7258-6416-4 (Hbk)
ISBN 978-3-7258-6417-1 (PDF)
https://doi.org/10.3390/books978-3-7258-6417-1

Contents

About the Editor

Mohammad Majid al-Rifaie

Mohammad Majid al-Rifaie is an Associate Professor in Artificial Intelligence at the University of Greenwich, School of Computing and Mathematical Sciences. He holds a PhD in Artificial Intelligence and Computational Swarm Intelligence (CSI) from Goldsmiths, University of London, and since the start of his PhD, he has published extensively in the field, covering various applications of CSI, Evolutionary Computation (EC), Machine Learning (ML) and Deep Neural Networks (DNNs). He has taught in the higher education for more than a decade, several of which relevant to CSI, AI, their real-world applications as well as philosophical issues on artificial intelligence and arts. His work in the area has featured multiple times in the media including the British Broadcasting Corporation (BBC) TV and Radio, the Guardian, and the New Scientist Magazine.

Over the past 10 years, he has developed a unique interdisciplinary research profile with more than 90 peer-reviewed publications, including book chapters and journal and conference papers on CSI, EC, ML, and DNNs as well as their applications in medical imaging, data science, philosophy, and arts. In addition to speaking multiple human languages, he is also familiar with around a dozen machine languages. He has supervised many PhD students and continues to do so.

Preface

This reprint, Swarm Intelligence and Evolutionary Algorithms for Real World Applications, brings together advances in Swarm Intelligence and Evolutionary Algorithms, offering insight into how nature-inspired methods address complex real-world challenges. Motivated by the limitations of conventional techniques in large, noisy, and incomplete search spaces, this volume highlights the robustness, adaptability, and emergent capabilities of SI and EC. Its aim is to foster deeper understanding and stimulate scientific dialogue on their practical applications. The reprint is intended for PhD students, early-career researchers, and experienced academics seeking reliable, innovative tools for solving difficult optimisation and modeling problems across diverse domains.

Mohammad Majid al-Rifaie
Guest Editor

algorithms MDPI

Article

Multi-Objective Unsupervised Feature Selection and Cluster Based on Symbiotic Organism Search

Abbas Fadhil Jasim AL-Gburi [1,*], Mohd Zakree Ahmad Nazri [1], Mohd Ridzwan Bin Yaakub [1] and Zaid Abdi Alkareem Alyasseri [2,3]

1 Faculty of Information Science & Technology, Universiti Kebangsaan Malaysia, Bangi 43600, Selangor, Malaysia; zakree@ukm.edu.my (M.Z.A.N.); ridzwanyaakub@ukm.edu.my (M.R.B.Y.)
2 Information Technology Research and Development Center (ITRDC), University of Kufa, Najaf 54001, Iraq; zaid.alyasseri@uokufa.edu.iq
3 College of Engineering, University of Warith Al-Anbiyaa, Karbala 56001, Iraq
* Correspondence: p104340@siswa.ukm.edu.my

Abstract: Unsupervised learning is a type of machine learning that learns from data without human supervision. Unsupervised feature selection (UFS) is crucial in data analytics, which plays a vital role in enhancing the quality of results and reducing computational complexity in huge feature spaces. The UFS problem has been addressed in several research efforts. Recent studies have witnessed a surge in innovative techniques like nature-inspired algorithms for clustering and UFS problems. However, very few studies consider the UFS problem as a multi-objective problem to find the optimal trade-off between the number of selected features and model accuracy. This paper proposes a multi-objective symbiotic organism search algorithm for unsupervised feature selection (SOSUFS) and a symbiotic organism search-based clustering (SOSC) algorithm to generate the optimal feature subset for more accurate clustering. The efficiency and robustness of the proposed algorithm are investigated on benchmark datasets. The SOSUFS method, combined with SOSC, demonstrated the highest f-measure, whereas the KHCluster method resulted in the lowest f-measure. SOSFS effectively reduced the number of features by more than half. The proposed symbiotic organisms search-based optimal unsupervised feature-selection (SOSUFS) method, along with search-based optimal clustering (SOSC), was identified as the top-performing clustering approach. Following this, the SOSUFS method demonstrated strong performance. In summary, this empirical study indicates that the proposed algorithm significantly surpasses state-of-the-art algorithms in both efficiency and effectiveness. Unsupervised learning in artificial intelligence involves machine-learning techniques that learn from data without human supervision. Unlike supervised learning, unsupervised machine-learning models work with unlabeled data to uncover patterns and insights independently, without explicit guidance or instruction.

Keywords: unsupervised learning; symbiotic organisms search algorithm; clustering; unsupervised feature selection; multi-objective

1. Introduction

In the ever-expanding landscape of data-driven applications, unsupervised learning techniques play a pivotal role in extracting meaningful patterns from raw data. Clustering, as one of the fundamental tasks in unsupervised learning, seeks to group similar data points together while maintaining separation between distinct clusters [1]. Several research works have been carried out, and various clustering approaches have been proposed, including kernel methods such as support vector machine (SVM) [2], self-organizing maps (SOM) [3], and k-means clustering [4]. However, achieving optimal clustering results remains a challenging endeavor due to various factors such as noisy features, high dimensionality, and the need for robust initialization.

Unsupervised feature selection (UFS), on the other hand, aims to identify the most relevant subset of features from the original feature space [5]. By selecting informative features, computational complexity would be reduced, and the quality of clustering results could also be enhanced. Based on evaluation criteria, the current unsupervised feature-selection studies can also be categorized into two primary groups: wrapper and filter-based studies [6]. The evaluation criterion for wrapper-based techniques is the chosen features' classification performance. On the other hand, the assessment criterion in filter techniques remains unaffected by the machine-learning technology. The filter approaches employ a variety of metrics, including distance measurements [7], consistency measures [8], correlation measures [9], and information theory-based measures [10]. Wrapper methods generally outperform filter methods because they evaluate the performance of the unsupervised selected features on a classification algorithm, even though filter methods are usually less computationally expensive [11]. However, these selection techniques continue to face issues with high computational time and convergence to local optima [12]. In addition, traditional unsupervised feature-selection methods often operate independently of the clustering algorithm, overlooking the inherent synergy between feature selection and the clustering process [13]. Metaheuristic techniques have been frequently adopted recently due to their robust global-search capabilities, which help overcome these shortcomings, especially when the number of features increases. Some of these classic metaheuristic algorithms, most widely applied to the unsupervised feature-selection and clustering problems, include the genetic algorithm (GA) [14–16], particle swarm optimization (PSO) [17–19], and harmony search algorithms, among others [20,21].

The SOS technique, first introduced by [22], is a stochastic metaheuristic approach using randomization to determine a collection of solutions. Based on the interactions between species in an ecosystem, the SOS algorithm was designed with a faster convergence time and greater robustness than these classic metaheuristic algorithms [23]. When compared to other population-based metaheuristic algorithms that searched for near-optimal solutions by training a set of candidate solutions using population characteristics to iteratively guide the searching, like the ant colony optimization (ACO) algorithm, the SOS algorithm is better for three key reasons. One benefit of the mutualism and commensalism stages of the SOS algorithm is that it concentrates on creating new creatures, which makes it possible for the algorithm to find a variety of solutions. It follows that the algorithm becomes more adept at exploring. For a second reason, the parasitism phase makes the algorithm more exploitation-capable by keeping it from being stuck in local optima. The final benefit is that there are just two general parameters for the SOS algorithm: the maximum number of iterations and the population size. Because of all these benefits, the SOS algorithm is widely used and has been modified to address a variety of optimization issues across a number of industries. Recently, to enhance its performance, modified [24] and hybrid [25] versions of the SOS algorithm have been developed as an alternative to the initial SOS algorithm proposed by [22]. Ref. [20] addressed the supervised feature-selection issue for 19 datasets from the UCI repository using the binary version of the SOS algorithm. The results indicated that, for the majority of datasets, the binary SOS algorithm may achieve a high classification accuracy with the fewest characteristics. The SOS algorithm has also been used to solve multi-objective problems in optimization. A multi-objective symbiotic organism search technique based on the weighted-sum method was proposed by [26] as a supervised learning method for economic/emission dispatch problems in power systems. The proposed method has been found to outperform other optimization algorithms such as the genetic algorithm (GA), differential evolution (DE), particle swarm optimization (PSO), the bees algorithm (BA), the mine blast algorithm (MBA), ant colony optimization (ACO), and cuckoo search (CS).

The application of SOS algorithms has since increased, particularly in the engineering field [27]. Though unsupervised learning has the capability to improve computational efficiency and retrieval recall, very few studies has been carried out in the literature specifically addressing unsupervised learning problems such as feature selection and clustering [27,28]. Previous studies concentrated on identifying optimal feature selection for brain–computer interfaces [26] and satellite image classification issues [29]. Within the literature, Refs. [30,31] explored text clustering and feature selection utilizing the SOS method. In their empirical research, Ref. [32] addressed text classification problems, and [31] proposed an SOS-based approach for feature extraction issues. Though the literature provides a larger proportion of works on single-objective approaches than on multi-objective optimization methods, it is observed that multi-objective optimization methods for FS problems based on metaheuristics techniques are not sufficiently examined [11]. Non-dominated sorting GA II (NSGA-II) or its variants form the basis of the majority of multi-objective techniques [33–36]. Other evolutionary computation (EC) approaches used in multi-objective feature selection include DE [37], ACO [38], and PSO [39]. According to the results of all these studies, multi-objective optimization algorithms outperform single-objective techniques in terms of both the quantity of features needed for supervised learning and classification performance. However, the existing literature has predominantly focused on datasets of modest to intermediate scale, indicating that the multi-objective feature-selection problem remains an unexplored field of study for unsupervised learning, much like high-dimensional data clustering. In addition, given that multi-objective evolutionary computation algorithms used for the FS problem are based on conventional algorithms like ACO, PSO, and GA, which typically have significant drawbacks such as slow convergence rate, high computational complexity, and trapping into local optima, there is also a need to investigate a novel multi-objective algorithm's capability to handle the feature-selection problems [33,40].

Although feature selection has been studied extensively, the literature review indicates that multi-objective unsupervised learning for two problems—unsupervised feature selection and clustering—has received relatively less attention. Furthermore, existing multi-objective research faces many of the aforementioned issues and has not addressed large-scale datasets such as TDT TREC data. This study proposes a multi-objective algorithm with a wrapper-based approach for data clustering, taking into account the shortcomings of the existing literature and the benefits of the SOS method. To the best of our knowledge, this study is the first to employ a multi-objective SOS algorithm to find the best possible unsupervised feature combination that maximizes clustering performance while decreasing the total number of selected features for a given set of data. To show the suggested method's robustness and dependability, it is evaluated using popular datasets from benchmark datasets. The results obtained are compared with the current approaches for both datasets, and the contribution related to the solution quality is given. The results of the study demonstrated that the proposed method performed better in terms of its capacity to provide acceptable outcomes, which included both an improvement in clustering performance and a reduction in the number of selected features. The robustness of this method is demonstrated by the better results it yields for both datasets. This work also examines and applies many SOS algorithm variants. The findings of these algorithms are compared with one another, and their benefits and drawbacks are identified.

The rest of the paper is organized as follows: Section 2 presents a review of related works, covering the background of the SOS algorithm, global-search unsupervised feature-selection algorithms based on SOS methods, and the clustering algorithm utilizing SOS algorithms. Section 3 outlines the proposed methods for this study. Sections 4 and 5 detail the experimental settings and results, respectively. Finally, the conclusion of the work is provided in Section 6.

2. Review of Related Works

This section includes the background of the SOS algorithm, global-search unsupervised feature-selection algorithms based on SOS methods, and clustering algorithms utilizing SOS approaches

2.1. Background of the SOS Algorithm

In this section, we introduce the original SOS algorithm, which is inspired by the three coexistence relations of mutualism, commensalism, and parasitism among organisms in the ecosystem. What follows explains these three phases of the SOS algorithm in detail.

2.1.1. Mutualism Phase

In this phase, interactions between two organisms, X_i and X_j, happen at random. Note that X_i and X_j, are two different organisms (where $X_i \neq X_j$). A mutual advantageous relationship between the two entities is formed by these interactions. Improved reciprocal survival rates of the two entities in the ecosystem are the goal of the correlation between X_i and X_j. By using Equations (1) and (2), the potential solutions X_{inew} and X_{jnew}, respectively, are obtained.

$$X_{inew} = X_i + \text{rand}(0,1) \times (X_{best} - X_{mutual} \times B_{F1}) \tag{1}$$

$$X_{jnew} = X_j + \text{rand}(0,1) \times (X_{best} - X_{mutual} \times B_{F2}) \tag{2}$$

where X_{mutual} is represented in Equation (3).

$$X_{mutual} = \frac{X_i + X_j}{2} \tag{3}$$

$$B_{F1} = 1 + \text{round}(\text{rand}(0,1)) \tag{4}$$

$$B_{F2} = 1 + \text{round}(\text{rand}(0,1)) \tag{5}$$

Within the range of 0 to 1, the $\text{rand}(0,1)$ function produces a vector of random numbers with a uniform distribution. According to [41], the organism designated as X_{best} exhibits the highest fitness function values in relation to their environmental adaptation. The term 'X_{mutual}', on the other hand, suggests that the two species exhibit mutualistic traits that promote their mutual survival. Equations (4) and (5) specify the random selection procedure used to determine the values of the benefit factors B_{F1} and B_{F2}. Those parameters show the degree of benefit resulting from interacting with each organism. Subsequently, the newly computed value of the fitness function is expressed as $f(X_{inew})$ and $f(X_{jnew})$. These values show better performance than the previous fitness functions, $f(X_i)$ and $f(X_j)$ [42]. Therefore, Equations (1) and (2) can be further transformed as follows:

$$X_{inew} = X_i + \text{rand}(0,1) \times (X_{best} - X_{mutual} \times B_{F1}) \text{if } f(X_{inew}) > f(X_i) \tag{6}$$

$$X_{jnew} = X_j + \text{rand}(0,1) \times (X_{best} - X_{mutual} \times B_{F2}) \text{if } f(X_{jnew}) > f(X_j) \tag{7}$$

2.1.2. Commensalism Phase

In the commensalism phase, two organisms, X_i and X_j, are randomly chosen from the ecosystem, and the organism X_i is updated according to Equation (8) [43].

$$X_{inew} = X_i + \text{rand}(-1,1) * (X_{best} - X_j), \text{ if } f(X_{inew}) > f(X_i) \tag{8}$$

A vector with randomly distributed values, which are evenly spaced over the range of -1 to 1, is produced by the $\text{rand}(-1,1)$ function. If organism X_{inew} shows a superior fitness value, it may replace organism X_i. ($X_{best} - X_j$) refers to the benefits that X_i gains with respect to X_j. In addition, there is no new solution for X_j, since X_j gains nothing from the interaction. In this phase, solution X_j serves to update X_i which is contrary to the mutualism phase, and therefore, the vector X_{inew} is updated according to Equation (8).

2.1.3. Parasitism Phase

In a symbiotic relationship where one species exclusively benefits at the expense of another, parasitism is exemplified by the interaction between humans, the Anopheles mosquito, and the Plasmodium parasite. The Anopheles mosquito, acting as a vector for the parasite, remains unaffected, while the human host experiences negative effects. According to [42], the Plasmodium parasite reproduces within the human body. Therefore, in the solution search space, organism X_i generates an artificial vector $X_{parasite}$ to mimic the parasitic behaviors previously reported for the Anopheles mosquito. This is accomplished by modifying organism X_i's randomly selected dimension through a process of adjustment [42]. Following that, an organism called X_j is randomly selected from the environment to serve as the host for the parasite $X_{parasite}$. The $X_{parasite}$ will then try to replace X_j within the ecosystem. If $X_{parasite}$ proves to be more fit than X_j, X_j will be replaced with $X_{parasite}$. This implies that X_j develops immunity to $X_{parasite}$, which ultimately causes $X_{parasite}$ to become extinct in the environment. This can be expressed as follows:

$$X_{parasite} = \text{rand}(0,1) \times (\text{UB} - \text{LB}) + \text{LB} \quad (9)$$

where the two boundary points that require addressing are denoted by LB (lower bound) and UB (upper bound). The majority of improvements to the traditional SOS algorithm have been achieved by the alterations to either the commensalism phase, the mutualism phase, or a combination of the two approaches. The addition of a fourth phase to the existing phases occurs only in exceptional and rare situations. This work presents a thorough analysis of several recent advances and hybridization approaches used in SOS algorithms, as reported in the previous studies.

2.2. Global-Search Unsupervised Feature-Selection Algorithms Based on SOS Methods

In the context of data mining and machine learning, unsupervised feature-selection methods have attracted a number of research interests as a result of their capability to identify and select relevant features without relying on class-label information. Feature selection aims to identify the relevant characteristics and remove the irrelevant features from the dataset in order to achieve benefits such as a decrease in the dimensionality of the data, an improvement in the performance of classification methods, and helps in the learning process [26]. However, it is noteworthy that there are two types of feature extraction approaches: wrapper-/coating-based and filter-based methods [21]. Coating-based methods use classification algorithms as a criterion for selecting the best possible solutions. Conversely, filter-based methods show a reasonable level of computing efficiency and are not dependent on any particular algorithm.

Researchers argue that since coating-based systems incorporate classification algorithms in the evaluation criteria, they perform better than filter-based methods [44,45]. In light of this background, the authors [21] offer different wrapper-based binary techniques that use the SOS method to solve feature-selection issues. BSOSST was first utilized, while BSOSVT was used as the second model. The binary SOS (BSOS) is developed using these two coating-based approaches. In the third method, EEBCSOS demonstrated its efficacy and exploratory capabilities. In order to address the problem of feature selection, Ref. [46] presented a novel technique called the improved binary symbiotic organism search (IBSOS), which makes use of the wrapper method. To preserve the delicate balance between exploration and exploitation, the authors also included the same biological symbiosis approaches that are used in the continuous SOS method in the proposed IBSOS technique. The NSMOSOS algorithm, a wrapper-based multi-objective algorithm, was introduced by [26] to generate the ideal feature subset. The study evaluated a brain–computer interface (BCI) system's effectiveness and robustness for motor-imaging feature selection across two datasets, achieving the highest accuracy results for both. In a recent study, Ref. [31] introduced an innovative feature-selection technique to extract the most relevant features from extensive input data, using the BSOS metaheuristic to enhance email spam detection. Research by [20] demonstrated that the BSOS algorithm could identify the minimal set of features across various datasets while maintaining high classification accuracy. However, the Bayesian structural optimization method (BSOS) has limitations with low-dimensional datasets and reduced sensitivity in high-dimensional datasets. Additionally, Ref. [47] proposed a new feature-selection approach employing the SOS method to enhance the accuracy and effectiveness of sleep staging by using physiological data to classify sleep stages. To cope with high-dimensional feature-selection problems, parallel multi-objective optimization approaches were also proposed by [48]. The proposed multi-objective evolutionary alternative (MOEA) approaches were implemented on EEG signals for brain–computer interface (BCI) benchmarks, which show the superior results in terms of hypervolume and speedup.

When comparing performance, wrapper methods generally surpass filter methods. However, for high-dimensional datasets like microarray datasets, where sample sizes are smaller than the feature dimensions, the wrapper method can become computationally intensive. To address this, researchers have developed hybrid strategies that integrate both wrapper and filter techniques [48,49]. The authors tested five distinct discrete SOS algorithms [48]. These techniques aim to improve classification accuracy by optimizing the neighborhood sizes of the k-nearest neighbor method and feature subsets through a two-step process. First, a large number of features are removed using a filter approach. Subsequently, a wrapper approach is applied to select the best subset among the other features that remains. Different approaches to SOS and feature selection are compiled in Table 1.

Table 1. A list of the various feature-selection techniques and SOS algorithms.

S/N	SOS Variations	Feature-Selection Approaches	References	Supervised or Unsupervised Learning
1	Modified SOS	Wrapped based	[20]	Unsupervised learning
2	Hybrid method	Wrapper/coating-based	[21]	Unsupervised learning
3	Multi-objective SOS	Wrapper based	[26]	Unsupervised learning
4	Modified SOS	Wrapper based	[31]	Unsupervised learning
5	Improved SOS	Wrapper based	[46]	Supervised learning
6	Modified SOS	Filter based	[47]	Unsupervised learning
7	MOEA	Wrapper based	[48]	Unsupervised learning
8	Five distinct SOS algorithms that combine modified and hybridized techniques	Wrapper based	[49]	Supervised learning
9	Hybrid approach	Filter and wrapper-based	[50]	Supervised learning
10	Hybrid method	-	[51]	Unsupervised learning
11	Hybrid method		[52]	Supervised learning

2.3. Clustering Algorithms Based on SOS Methods

According to [26,49,53], SOS algorithms have been widely used for classification problems but not yet evaluated for the unsupervised feature-selection and clustering problems. An unsupervised technique for analyzing data referred to as clustering is used to find sets of objects that are homogeneous based on the values of their properties. Clustering can be divided into two primary categories: partitional clustering methods and hierarchical clustering methods. The recurrent hierarchical grouping of data objects is an essential component of the hierarchical clustering technique. In addition, a single partition of a dataset is created using the partitional clustering technique to find the underlying categories in the data. This method is non-hierarchical and is based on the application of a predetermined objective function. k-means is an example of this.

In an effort to further solve problems associated with automatic data clustering, an automatic k-means clustering-based SOS (CSOS) was proposed by [54]. By integrating the global and local search strategies in a sub-ecosystem built on the automated k-means clustering technique, the CSOS algorithm essentially performs a hybrid search approach. Mutualism and commensalism are the two distinct phases that are covered by the local search. Furthermore, in CSOS global search, this is referred to as the parasitism phase, in which only the best solution within a cluster can communicate with the best solutions in other clusters. Additionally, a data categorization methodology combining an SOS algorithm and a regularized extreme learning machine was developed by [55]. The newly developed algorithm, SOS-RELM, consists of two separate SVM phases: backpropagation and least-squares support vector machines. As a result, the SOS method is both an effective and efficient optimization approach. In the second phase, it optimizes the regularization parameters, invisible biases, and input weights. Moreover, [49] presented DSOS algorithms in a different study for the optimization of neighborhood size and feature subsets as supervised learning.

Furthermore, [53] presented a method for automatic data clustering that makes use of the SOS algorithm. In this article, we have discussed the ease of trapping in a local optimum and the strong dependence of the k-means clustering algorithm on the beginning solution. To create clusters, the SOS framework used an automated k-means clustering algorithm. The most effective solutions interact with one another within each cluster, combining local and global search techniques [56]. However, such an approach leads to a rise in computational cost.

Similarly, by including a multi-agent system (MAS) and self-adaptive benefit factors in the SOS algorithm, [57] created a multi-agent SOS (MASOS) to enhance the performance of the original SOS algorithm. Using this method, every organism acts as an agent interacting locally to choose the best course of action. When an agent chooses another agent from its immediate area, the SOS algorithm would execute in three separate steps. Despite that, there was a lot of computational complexity in the proposed task. The adopted clustering approaches that were described are listed in Table 2.

Table 2. The clustering approach adopted.

Authors	Adopted Clustering Approach
[54]	The number of clusters that form at the beginning is half of the ecosize, which is divided into smaller ecologies. Subsequently, CSOS optimization is then applied.
[58]	Enhancing the BDI and BIC through the optimization process.
[56]	The CVI functions as the objective function in the optimization problems of the Davies–Boulding index and the compact separated index, both of which must be minimized.
[53]	To optimize the clustering problem, the SOS algorithm randomly initializes the cluster within the ecosystem.
[59]	The benefit factors are found by a non-linear method, and their weights are used to effectively explore and exploit the search region.
[60]	The number of initial clusters is determined by the eco-size, which is the number of sub-ecosystems that the self-organizing system (SOS) forms and then optimizes.

3. Proposed Method

Unsupervised feature selection plays a crucial role in mitigating the curse of dimensionality, particularly in tasks like document clustering where high-dimensional text data are involved. The role of feature selections is multifaceted. They serve purposes such as enhancing performance (e.g., accuracy), aiding data visualization, simplifying model selection, and minimizing dimensionality by eliminating noise and unnecessary attributes [61,62].

In this study, a symbiotic organism search algorithm (SOS) was developed to solve numerical optimization over a continuous search space. The proposed SOS algorithm, like other population-based methods, iteratively employs a population of candidate solutions within the search space to find the optimal global solution. SOS starts with an initial population (referred to as the ecosystem), where organisms are randomly generated. Each organism represents a candidate solution, and its associated fitness value reflects its adaptation to the desired objective. This approach models the ecological interaction between two organisms in the ecosystem to control the production of new solutions. The three phases, including parasitism, commensalism, and mutualism, which resemble the real-world biological interaction framework, are shown.

The nature of interactions determines the primary principle for each phase. In the mutualism phase, interactions benefit both sides. In the commensalism phase, one side benefits without affecting the other. In the parasitism phase, one side benefits while actively harming the other. Throughout all stages, interactions between the organisms are random and continue until the termination conditions are met. The SOS algorithm processes are described in full in the following algorithm, and further information about the three phases as provided in the next section include

Initialization
REPEAT

1. Mutualism phase.
2. Commensalism phase.
3. Parasitism phase.

UNTIL (the termination criterion is met).

3.1. Mutualism Phase

An illustrative example of mutualism, which benefits both participating organisms, is the symbiotic relationship between bees and flowers. Bees actively fly among flowers, collecting nectar that they transform into honey—a process beneficial to the bees themselves. Simultaneously, this activity also benefits the flowers, as bees inadvertently distribute pollen during their foraging, facilitating pollination. In the context of the SOS phase, this mutualistic interaction serves as a model.

In SOS, X_i is an organism matched to the ith member of the ecosystem. Another organism X_j is then selected randomly from the ecosystem to interact with X_i. Both organisms engage in a mutualistic relationship to increase the mutual survival advantage in the ecosystem. New candidate solutions for X_i and X_j are calculated based on the mutualistic symbiosis between organism X_i and X_j, which is modeled in Equations (10) and (11).

$$X_{inew} = X_i + \text{rand}(0,1)\times\ (X_{best}\text{-mutual vector} \times \text{BF}_1\ ), \tag{10}$$

$$X_{jnew} = X_i + \text{rand}(0,1)\times\ (X_{best}\text{-mutual vector} \times \text{BF}_2), \tag{11}$$

$$\text{Mutual Vector} = \frac{X_i + X_j}{2} \tag{12}$$

rand(0,1) in Equations (10) and (11) is a vector of random numbers.

What follows explains the function of BF1 and BF2. In the natural world, certain mutualistic connections may benefit one organism more than another. In another context, interactions with organism B may be extremely advantageous for organism A. When interacting with organism A, organism B may only receive minimal or insignificant benefits. In this case, benefit factors (BF_1 and BF_2) are arbitrarily assigned to 1 or 2. These variables indicate the extent to which each organism benefits from the contact—that is, whether one organism gains all or some benefit from it.

The relationship feature between organisms X_i and X_j is represented by a vector named 'Mutual_Vector', as shown in Equation (12). The mutualistic effort to accomplish their objective of enhancing survival advantage is reflected in the (X_{best}-Mutual Vector $\times$ BF_1) component of the equation. Moreover, all species are compelled to increase their degree of adaptation to their ecosystem, based on Darwin's theory of evolution, which states that 'only the fittest organisms will prevail'. Some of them enhance their adaption to survival by forming symbiotic relationships with other organisms. Since X_{best} represents the maximum level of adaptation, it is required in this scenario. Consequently, we model the highest degree of adaptation as the objective point for the fitness increment of both organisms using (X_{best}/global solution). In the end, organisms are only updated if their current fitness exceeds their fitness before the interaction.

3.2. Commensalism Phase

A common example of the interaction between remora fish and sharks can be used to define commensalism. In such a scenario, the remora gains the advantage when it clings to the shark and consumes its remaining food. The behaviors of remora fish do not affect the shark, and their relationship offers very little benefit to it.

As in the mutualism phase, an organism denoted X_j is chosen at random from the environment to engage in interactions with X_i. In this situation, organism X_i makes an effort to gain something from the exchange. However, the relationship does not benefit or harm organism X_j itself. The commensal symbiosis between organisms X_i and X_j, which is described in Equation (13), is used to determine the new candidate solution of X_i. By the rules, organism X_i is modified only if its current fitness exceeds its fitness before the interaction.

$$X_{inew} = X_i + \text{rand}(-1, 1) \times X_{best} - X_j \tag{13}$$

The portion of the equation denoted by ($X_{best} - X_j$) reflects a positive advantage that X_j offers by helping X_i maximize its survival advantage within ecosystems in the current organism (represented by X_{best}).

3.3. Parasitism Phase

The Plasmodium parasite, which spreads between human hosts using its connection with the Anopheles mosquito, is a good example of parasitism. Whereas the parasites grow and replicate in the human body, their human host may suffer malaria and die as a result. By creating a synthetic parasite known as 'Parasite Vector', SOS gives organism X_i a function like that of the Anopheles mosquito. Using a random value to adjust the randomly chosen dimensions, organism X_i is duplicated in the search space to form a parasite vector. The host for the parasite vector is an organism denoted by X_j, which is chosen at random from the ecosystem. X_j is being replaced by the parasite vector in the ecosystem. Consequently, the fitness of both organisms is determined.

If the parasite vector has higher fitness values than organism X_j, it will be eliminated and its place in the ecosystem will be taken over. In other words, if the fitness value of X_j is higher, the parasite will not be able to survive in that ecosystem since X_j will be resistant to it.

3.4. Development of Initial Features

The values of selected features appear to be organized as an array. In optimization terminology, particle position corresponds to this array in particle swarm optimization (PSO), while genetic algorithms refer to it as a 'Chromosome'. As a result, the proposed approach labels each individual feature as a 'Raindrop' feature. In the problem selection of N_{var} dimensional features, a raindrop represents an array of $1 \times N_{var}$. Such an array is explained as follows:

$$\text{Feature of symbiotic} = [X_1, X_2, X_3 \ldots X_N] \tag{14}$$

At the beginning of the feature selection, a candidate representative of a matrix of size raindrops $N_{pop} \times N_{var}$ is created (i.e., features raindrops). Then, the matrix X is randomly created and provided as follows (columns and rows are the quantity of the variable of design and the quantity of unsupervised feature selections):

$$\text{Feature Ecosystem} = \begin{bmatrix} \text{eco}_1 \\ \text{eco}_2 \\ \text{eco}_3 \\ \vdots \\ \text{eco}_{\text{Neco\_size}} \end{bmatrix} \quad (15)$$

$$\begin{bmatrix} x_1^1 x_2^1 x_3^1 & \cdots & x_{\text{Nfeature}}^1 \\ \vdots & \ddots & \vdots \\ x_1^{\text{Neco\_size}} x_2^{\text{Neco\_size}} x_3^{\text{Neco\_size}} & \cdots & x_{\text{Nfeature}}^{\text{Neco\_size}} \end{bmatrix}$$

Every value of the decision variable (X_1, X_2, $X_3 \ldots X_{Nvar}$) can be described as the following numbers (0 or 1), where N_{vars} and N_{pop} are the number of design variables and the number of raindrops (preliminary unsupervised features selection), respectively. Moreover, N_{pop} raindrops are generated, and subsequently, the raindrop cost is obtained by the assessment of the function of cost (*Cost*) as follows.

$$\text{Cost}_i = f\left(x_1^i, x_2^i, \ldots x_{\text{Nfeature}}^i\right) \; i = 1,\ 2,\ 3, \ldots,\ \text{Neco\_size}. \quad (16)$$

3.5. Cost of Solutions

As previously established, each row in eco is associated with many features in the document. In the context of eco a row's set of features is represented is denoted by f = (f_1, $f_2 \ldots f_k$). The objective for each row in eco is to assess the mean absolute difference (MAD), as detailed in [61]. MAD is used to determine the most relevant features for text classification by correlating the scores with the importance of each feature. The aim is to assign scores that accurately reflect the significance of each feature. One way to obtain such a score is to take the difference between the mean values and the sample. It can be depicted as per the following equation:

$$MAD_i = \frac{1}{n_i} \sum_{j=1}^{n} \left| X_{ij} - \overline{X_i} \right| \quad (17)$$

where X_{ij} is the value of feature *i* in accordance with the document *j* and X_i is the mean of the feature *i*, which is computed according to the equation as follows:

$$\overline{X_i} = \left(\frac{1}{n}\right) \sum_{j=1}^{n} X_{ij} \quad (18)$$

Every element in the solution indicates the cluster number as C = (c_1, $c_2 \ldots c_k$), and the computation of each solution in eco corresponds to a document cluster. The set of *K* centroids that correspond to a row in eco is represented by the C. The centroid of the *k*th cluster can be calculated as follows: $c_k = (c_{k1} \ldots,)$.

$$c_{kj} = \frac{\sum_{i=1}^{n} a_{ki} d_{ij}}{\sum_{i=1}^{n} a_{ki}} \quad (19)$$

The goal is to verify that by minimizing distances within and between clusters, cluster centroids optimize similarity both within and between clusters. The row corresponds to the average distance of documents from the cluster centroid and the associated fitness value. Following this, a suitable solution is derived based on this information. The condition is commonly known as attention deficit disorder with hyperactivity (ADDC).

$$Cos_i = \left[\sum_{i=1}^{K} \frac{1}{ni} \sum_{j=1}^{n} D(Ccent,\ dj)\right] / K \quad (20)$$

The cosine similarity is denoted by the D (.,.), where d_{ij} is the *j*th document in cluster *i*, *K* represents the number of clusters, and n_i is the number of documents contained in cluster i (e.g., ($n_i = \sum_{j=1}^{n} a_{ij}$)). If the cost value of the locally optimized vector is higher than that of the eco solutions, the newly solution generated can be replaced within a row in eco.

4. Experimental Settings

4.1. Parameter Setting of Symbiotic Organisms Search as Unsupervised Feature Selection

This section examines how the solutions of the algorithms evolved over generations under various configurations of just one important parameter. This is eco, where eco is the organism count (the initial feature population). In this case, this section clarifies the effects of changing certain parameters and specifically looks at three different scenarios, as shown in Table 3. Furthermore, utilizing the internal evaluation known as MAD, the experimental investigation demonstrated that the best results were obtained by a clear relationship between ecoNeco_size and the number of features. We looked at every situation and determined that the maximum repetition count for each run should be 100. Section 3.5 discusses the use of MAD to determine the fitness function value, which is the solution cost value. Additionally, the unsupervised symbiotic organism search algorithm used for the assessment is based on the feature selection covered in Section 2.2, at $d_{max} = 1 \times 10^3$. The scenario with ecoNeco_size = 40 is the best one, related to the fifth.

Table 3. Some scenarios of the parameters' symbiotic organisms search as the feature selection.

Scenarios	eco_{Neco_size}
1	8
2	16
3	24
4	32
5	40
6	48
7	56
8	64
9	72

4.2. Investigating the Impact of Different SOS Parameters on Cluster SOS

The aim of this section is to examine how the algorithms' solutions have evolved over generations under various configurations for a single parameter. This is ecoNeco_size, where ecoNeco_size is the quantity of documents and the number of groups, both of which are user-specified.

Given these conditions, this section focuses on highlighting the effect of a single parameter changes. Specifically, the following three scenarios have been tested and are displayed in Table 4. Empirical research has also demonstrated that the best results can be obtained with a linear relationship between ecoNeco_size and the number of clusters. All scenarios were tested over ten runs, with a maximum number of iterations fixed at 100 for all runs. The fitness function value is the ADDC value of the solution. The SOS algorithm, which is covered in Section 5, is the evaluation algorithm. $d_{max} = 1 \times 10^3$ was obtained using the Routers dataset.

The evolution of the solution for various SOS parameter values as clusters shows that when ecoNeco_size is decreased, the solution is found more slowly than when ecoNeco_size is increased, which causes the solution to be found faster than when ecoNeco_size is small. Nevertheless, with the benefits of reducing the required amount of space and converging to the optimal result, using the appropriate ecoNeco_size seems to be a reasonable and logical choice. In addition, doubling the number of clusters (8 × 2 in this dataset) can yield the best results with a specific selection of ecoNeco_size.

Table 4. Scenarios for analysis of SOS convergence behavior.

Scenario	Npop
1	2
2	4
3	6
4	8
5	16
6	18
7	20
8	24
9	28

Performance Measurement and Datasets

The universal F-measure from [63] was used in this study to evaluate the external condition. The F-measure, which takes into account recall and accuracy of information retrieval, is high if clusters are ideal. Each class is expected to have its own set of essential documents, and each class will continue to have its required document compilation. Higher F-measure values indicate better clustering performance, with a metric comprising a range of 0 to 1. Equation (21) shows the mathematical expression of F-measures for cluster j and class i:

$$\mathrm{F}(i,j) = \frac{2\mathrm{Recall}(i,j) \times \mathrm{Precision}(i,j)}{\mathrm{Recall}(i,j) + \mathrm{Precision}(i,j)}. \tag{21}$$

Equation (22) is used to determine the overall value for the F-measure, which is taken as in the weighted average of all.

$$\mathrm{F} = \sum_i \frac{n_i}{\mathrm{N}} maxF(i,j) \tag{22}$$

Thus, the F-measure values are distributed across the interval (0,1), and the values are proportional to higher levels of clustering quality.

This study used four independent and separate datasets to conduct a thorough evaluation of algorithm performance. This approach guaranteed an exhaustive and objective evaluation and comparison of the algorithms. The main dataset, known as classic 3, was used as a benchmark for comparison during the text-mining process. There are 3892 documents in the collection, which have been classified into three groups. In particular, there are 1460 documents centered on information retrieval (CISI), 1033 documents referring to medical problems (MEDs), and 1399 documents related to aviation systems (CRANs) [64].

The second dataset was made up of 1445 CNN news articles that were selected from the TDT2 and TDT3 corpora. Replicating the dataset, the i-Event experiment uses information from the TDT2 and TDT3 corpora [65]. A significant number of chosen documents are usually required to experiment and create clusters on the user interface. The concision of CNN's reporting and the significance of the events they covered also had a role in the choice of these sources.

The 20 newsgroups data [66] are the data used in this study and include 10,000 messages in total. Each Usenet newsgroup had 1000 messages, and these messages were collected from ten different newsgroups. The current study utilizes this dataset as its third dataset, which had 3831 documents in total after pre-processing. Thus, to assess how effectively algorithms handle large-scale datasets, a dataset with 20 newsgroups was utilized.

Another popular dataset used extensively in earlier academic studies is Reuters-21578 [67], which is a test set for the classification of text. However, there are several limitations related to the procedure of data collections in this setting. Many documents belong to more than one class, and most of the papers do not have labels for the class annotations. Moreover, the dataset distribution shows consistency across different groups. Some classes, like 'earn' and 'acquisition', have a huge number of documents, while other classes, like 'reserve' and 'veg-oil', have very few documents. To overcome these constraints, this study used a data with eight main groups and 1100 documents in each group. The summary description of document set is given in Table 5 with number of documents and number of clusters respectively.

Table 5. Summary description of document set.

Document	Source	#of Document	#of Cluster
DS1	Classic 3	3892	3
DS2	TDT2 and TDT3 of TREC 2001	1445	53
DS3	20 NEWSGROUP	3831	10
DS4	routers	4195	8

represent the number of documents and clusters in each category.

5. Results and Discussion

5.1. Evaluation of the SOS Cluster Using All Features

In this section, we evaluate different algorithms, including harmony search as clustering (HSCLUST), k-means, one-step k-means with harmony search (KHS), and SOS, using standard datasets. For all these algorithms, the similarity metric was the cosine correlation measure. Notably, the results presented in this section represent the average performance over 20 runs (to ensure unbiased comparisons). Additionally, each algorithm is executed with 1000 iterations per run. It is worth noting that no specific parameters need to be set up for the k-means algorithm. For SOS, the ecoNeco_size is set to twice the number of classes in the dataset. This paper adapts the same settings as [61] Bsoul et al. (2021) for the other HSCLUST algorithm. Specifically, the authors set the HMS to be twice the number of clusters in the dataset, and PARmax was set to 0.9, PARmin was 0.45, and HMCR was set to 0.6.

In Table 6, the algorithmic performances in the document collections are evaluated based on the F-measure. Among the different algorithms, HS + k-means stands out with the highest F-measure. On the other hand, HS as clustering performs poorly. Interestingly, the proposed SOS algorithm is comparable to the k-means algorithm. Specifically, in four datasets, SOS outperforms k-means. In terms of clustering, SOS outperforms HS when compared to the SOS algorithm. As a result, SOS outperforms HS in locating the optimum centroids, while k-means is not very good at locating the global optimal initial centroids. In other words, the SOS is not as powerful in local optima as k-means. Building upon this observation, the next section explores the combination of SOS-based unsupervised feature selection and SOS in clustering. The goal in the subsequent sub-section is to leverage SOS's strength in finding global optima most important features.

Table 6. The result of three cluster algorithms using the external evaluation F-measure.

Datasets	k-Means	HS	SOS	KHS
Classic 3	0.929	0.909	**0.933**	<u>**0.935**</u>
TREC 2001	0.804	0.829	**0.831**	<u>**0.853**</u>
Newsgroup	0.582	0.611	**0.649**	<u>**0.659**</u>
Routers	0.636	0.682	**0.702**	<u>**0.712**</u>

Best result: underline and bold; second-best result: bold.

Evaluation of SOS-Based Unsupervised Feature Selection with SOS Cluster

As shown in Table 7, the summary comprises the number of features derived from the k-means and bag-of-words (BOW) models, as well as the results of the SOSFS and the SOSC algorithms for cluster performance evaluation. Based on our findings, DS1 had the best F-measure (92.9%) obtained by using the k-means clustering algorithm, whereas DS3 had the lowest F-measure (58.2%). For dataset DS1, the optimization of the PSOC method yielded the highest F-measure of 89.1%. In contrast, for dataset DS3, PSOC resulted in the lowest F-measure of 60.6%. When utilizing the HSC method, dataset DS1 achieved the highest F-measure of 90.9%, whereas dataset DS3 had the lowest F-measure of 61.4%. For the SOSC technique, dataset DS1 produced the highest F-measure of 93.3%, while dataset DS3 recorded the lowest F-measure of 64.9%. Finally, the KHCluster method achieved the maximum F-measure of 93.5% for dataset DS1, but for dataset DS3, KHCluster resulted in the lowest F-measure of 65.9%.

Table 7. The f-measurement of the cluster using all features.

Comparison	DS1 Classic 3		DS2 TREC 2001		DS3 Newsgroup		DS4 Routers	
	Features	F-Measure	Features	F-Measure	Features	F-Measure	Features	F-Measure
k-means	13,310	0.929	6737	0.804	27,211	0.582	12,152	0.636
PSOC	13,310	0.891	6737	0.841	27,211	0.606	12,152	0.688
HSC	13,310	0.909	6737	0.829	27,211	0.611	12,152	0.682
WDOC	13,310	**0.933**	6737	**0.83**	27,211	**0.64**	12,152	**0.69**
KHCluster	13,310	<u>**0.935**</u>	6737	<u>**0.853**</u>	27,211	<u>**0.659**</u>	12,152	<u>**0.712**</u>

Best result: underline and bold; second-best result: bold.

Table 8 illustrates the performance of the SOSFS algorithm with various clustering techniques:

Table 8. The f-measurement of the proposed SOS-based unsupervised feature selection.

Comparison	DS1 Classic 3		DS2 TREC 2001		DS3 Newsgroup		DS4 Routers	
	Features	F-Measure	Features	F-Measure	Features	F-Measure	Features	F-Measure
k-means	<u>**8186**</u>	0.93	5573	0.824	20,854	0.602	9561	0.636
PSOC	9927	0.928	<u>**4826**</u>	0.847	**13,824**	0.636	<u>**5084**</u>	0.688
HSC	10,843	0.929	**5128**	0.831	<u>**11,843**</u>	0.621	10,854	0.682
WDOC	**9824**	**0.939**	5834	**0.83**	19,283	**0.64**	6891	**0.69**
KHCluster	10,289	<u>**0.949**</u>	6057	<u>**0.861**</u>	20,851	<u>**0.668**</u>	**5732**	<u>**0.748**</u>

Best result: underline and bold; second-best result: bold.

For dataset DS1, the integration of SOSFS with k-means achieved the highest F-measure of 93%. Conversely, for dataset DS3, the same integration produced the lowest F-measure of 60.2%. The SOSFS algorithm effectively reduced the number of features in DS1 by 8186 out of 13,310, in DS2 by 5573 out of 6737, in DS3 by 20,854 out of 27,211, and in DS4 by 9561 out of 1215.When SOSFS was combined with PSO, the lowest F-measure of 63.6% was observed for dataset DS3. The feature reduction for SOSFS with PSO was significant: 9927 out of 13,310 features in DS1, 4826 out of 6737 in DS2, 13,824 out of 27,211 in DS3, and 5084 out of 12,152 in DS4.With SOSFS and HSC, dataset DS3 had the lowest F-measure of 62.1%. Feature reduction was also substantial: 10,843 out of 13,310 in DS1, 5128 out of 6737 in DS2, 11,843 out of 27,211 in DS3, and 108,54 out of 12,152 in DS4.

When the SOSFS algorithm was combined with WDOC, dataset DS3 yielded the lowest F-measure of 65%. In DS1, DS2, DS3, and DS4, SOSFS effectively decreased the number of features by 9824 out of 13,310, 5834 out of 6737, 19,283 out of 27,211, and 6891 out of 12,152. The lowest F-measure of 66.8% was achieved for dataset DS3 when the SOSFS algorithm was combined with the KHCluster. In DS1, DS2, DS3, and DS4, SOSFS effectively decreased the number of features by 10,289 out of 13,310, 6057 out of 6737, 20,851 out of 27,211, and 5732 out of 12,152. For SOSUF reduction in DS1, k-means was the best. In DS2, SOSUFS was best in PSOC; in DS3, it was best with HSC; and in DS4, it was best with PSOC. Comparing Tables 7 and 8, the KHCluster showed the best performance and was more powerful than used all the features. From Tables 7 and 8, we can observe that the SOSUFS proved that some of the features mis-cluster and reduce the performance of cluster algorithms. In all cluster algorithms, the SOSUFS enhances the performance when looking to the F-measure.

As seen in Table 9, when combined with SOSC, the SOSFS approach produced the highest F-measure for dataset DS1: 95.3%. In DS1, DS2 and DS3, 10,346 and 12,152 features, 7096 and 4731 features, and 5651 and 12,152 features, respectively, were successfully reduced by SOSFS. The proposed symbiotic organisms search-based optimal unsupervised feature selection (SOSUFS) in conjunction with symbiotic organism search-based optimal clustering (SOSC) was found to be the best text clustering strategy. Following the evaluation, the SOSFS algorithm combined with the k-means algorithm demonstrated a relatively strong performance. The SOSC and particle swarm optimization clustering (PSOC) algorithms also showed moderate performance. In contrast, the KHCluster method ranked third in terms of performance. Out of all the approaches that were assessed, the k-means algorithm performed the worst. Text clustering performance has been demonstrated to be enhanced by using the UFS with the k-means algorithm.

Table 9. The f-measurement of the proposed hybrid multi-objective clustering.

Comparison	DS1 Classic 3		DS2 TREC 2001		DS3 Newsgroup		DS4 Routers	
	Features	**F-Measure**	**Features**	**F-Measure**	**Features**	**F-Measure**	**Features**	**F-Measure**
k-means	**8186**	0.93	5573	0.824	20,854	0.602	9561	0.636
PSOC	9927	0.928	**4826**	0.847	13,824	0.636	**5084**	0.688
HSC	10,843	0.929	5128	0.831	**11,843**	0.621	10,854	0.682
WDOC	9824	0.939	5834	0.83	19,283	0.64	6891	0.69
KHCluster	10,289	**0.949**	6057	**0.861**	20,851	**0.668**	5732	**0.748**
SOSFS with SOSC	**7096**	**0.953**	**4731**	**0.865**	**10,346**	**0.686**	**5651**	**0.734**

Best result: underline and bold; second-best result: bold.

A statistical evaluation of the proposed hybridized SOS multi-objective methods for clustering, unsupervised feature selection was conducted. The optimal performance of text clustering can be assessed, as can whether significant differences can be obtained, using this multi-objective method for clustering and unsupervised feature selection. Based on the Friedman criterion, Table 10 shows the ranks of the multi-objectives proposed for the clustering, unsupervised feature selection and other methods. The criteria establish the rating, with a lower value signifying a higher rank.

Table 10. The ranking of the proposed algorithms using the Friedman test.

Algorithms	Ranking
k-means	10.18
PSOC	10.15
HSC	10.01
WDOC	9.61
SOSC	9.39
KHCluster	9.3
SOSC and SOSFS	**8.53**
Friedman test (p-value)	0.00
man-Davenport (p-value)	0.00

The best is in bold font.

Table 10 demonstrates that our unsupervised feature selection and grouping technique, which employs SOS, had the lowest value and was ranked highest. The last two rows of Table 10 display the Friedman and Iman–Davenport p-values. Hybridize SOS in unsupervised feature selection with SOSC, KHCluster, SOSCe, WDOC, HSC, PSOC, and k-means is in the second, third, fourth, fifth, sixth, and seventh ranks, respectively.

5.2. Discussion

This study focused on the problems of finding a near-optimal partition cluster concerning the ADDC criterion for a given set of texts, to split them into a specified number of clusters and find near-optimal features. To this point, this work studied each of the existing clustering methods and detected the behavior of k-means, HSC, KHSC, and PSO. Although each algorithm exhibits considerable performance, some of the weaknesses of each algorithm are found and discussed. The k-means algorithm, for instance, performed depending on the initial centroid selected. However, the memory requirements for large datasets are its major limitation. In addition, the k-means algorithm is also found to perform better than the PSO algorithm. Given the limitations of existing clustering methods, metaheuristics approaches, such as harmony search clustering, filled the gaps. The HS method shows an amazing performance compared to other approaches based on the ADDC evaluation, whereas HSCLUST was found to have the worst performance compared with other traditional clusters. The proposed algorithms using SOS in unsupervised feature selection and in clustering in this article aimed to address the limitations of the traditional methods by demonstrating the partitioning problems as optimization issues. Primarily, the symbiotic organism search algorithm, the SOS, was employed as an unsupervised feature-selection and cluster method to optimize the objective functions related with the optimal clustering method. The effect of the ecoNeco_size parameter was verified, and the experimental works show that with linear relations between ecoNeco_size and the number of features and clusters, better results would be possible. As shown from the results, ecoNeco_size has a better performance with two times the number of classes in the dataset. The first experimental findings revealed that the SOS method has equivalent performance to the k-means algorithm but performed better than the HS method for the clustering task. Based on the experimental analysis of benchmark datasets, the result shows that the hybridization of SOS-based unsupervised feature selection with SOS in clustering was better than other hybrid approaches, due to the effective nature of the SOS, which focuses on the global optima features and centroids. However, the results show that the performance was better achieved by combining the SOS-based unsupervised feature-selection method with an SOS-based cluster method compared to KHS, SOS, k-means, and HS, respectively.

6. Conclusions

This study investigates a multi-objective symbiotic organism search algorithm for unsupervised feature selection that simultaneously takes into account the number of selected features and the accuracy of the clustering. The effectiveness of the proposed approach is examined using reputable benchmark datasets. Based on the results, the proposed feature-selection model offered a feature subset of high quality for feature selection, and the final model outperformed the other techniques in terms of clustering accuracy for both datasets. The fact that the proposed feature-selection technique produced satisfactory results for both datasets indicates that it may be able to provide a potential solution to this problem. In addition, the robustness of the model across datasets is also demonstrated by this finding. However, in comparison to the traditional unsupervised feature-selection algorithms, the computing cost of the proposed algorithm is significant. Although unsupervised feature selection is typically an offline procedure, this is not a total disadvantage. Additionally, hybrid techniques are presented in this paper and applied to all datasets for unsupervised feature selection. These algorithms were shown to have both strengths and limitations. Out of all the algorithms discussed in this study, the combination of SOSFS and SOSC produced the best clustering accuracy results while reducing features that are not relevant. This approach presents an optimal solution set instead of a single solution and demonstrates the significance of handling feature-selection problems as a multi-objective process. As a result, this study concludes that the proposed SOSFS is a realistic and efficient method for solving unsupervised feature-selection problems. Additionally, the proposed SOSC model outperformed the other algorithms for both datasets in terms of clustering accuracy. Furthermore, various datasets for feature selection and clustering may also be used to test the proposed hybrid approach. In addition, future research needs to examine the nature of features chosen by the SOSFS algorithm when combine with k-means and SOSC, which have not yet been evaluated.

Author Contributions: Conceptualization, A.F.J.A.-G., M.Z.A.N. and Z.A.A.A.; methodology, A.F.J.A.-G., M.Z.A.N. and Z.A.A.A.; software, A.F.J.A.-G.; validation, A.F.J.A.-G., M.Z.A.N. and M.R.B.Y.; formal analysis, A.F.J.A.-G., M.Z.A.N. and M.R.B.Y.; investigation, A.F.J.A.-G. and M.Z.A.N.; resources, A.F.J.A.-G., M.Z.A.N. and Z.A.A.A.; data curation, A.F.J.A.-G.; writing—original draft preparation, A.F.J.A.-G., M.Z.A.N. and M.R.B.Y.; writing—review and editing, A.F.J.A.-G., M.Z.A.N. and Z.A.A.A.; visualization, A.F.J.A.-G.; supervision, M.Z.A.N. and M.R.B.Y.; project administration, Z.A.A.A.; funding acquisition, A.F.J.A.-G. All authors have read and agreed to the published version of the manuscript.

Funding: This research was funded by Malaysia Ministry of Higher Education and the Universiti Kebangsaan Malaysia under the Q2678Q25 FRGS/1/2022/ICT02/UKM/02/3 with machines and materials.

Data Availability Statement: The dataset used in this work based on [64,65].

Acknowledgments: We thank the anonymous reviewers for their comments and suggestions.

Conflicts of Interest: The authors declare no conflicts of interest.

References

1. Oyewole, G.J.; Thopil, G.A. Data clustering: Application and trends. *Artif. Intell. Rev.* **2022**, *56*, 6439–6475. [CrossRef] [PubMed]
2. Gedam, A.G.; Shikalpure, S.G. Direct kernel method for machine learning with support vector machine. In Proceedings of the 2017 International Conference on Intelligent Computing, Instrumentation and Control Technologies (ICICICT), Kerala, India, 6–7 July 2017; pp. 1772–1775.
3. da Silva, L.E.B.; Wunsch, D.C. An Information-Theoretic-Cluster Visualization for Self-Organizing Maps. *IEEE Trans. Neural Netw. Learn. Syst.* **2017**, *29*, 2595–2613. [CrossRef] [PubMed]
4. Sinaga, K.P.; Yang, M.-S. Unsupervised K-Means Clustering Algorithm. *IEEE Access* **2020**, *8*, 80716–80727. [CrossRef]
5. Wang, P.; Xue, B.; Liang, J.; Zhang, M. Feature clustering-Assisted feature selection with differential evolution. *Pattern Recognit.* **2023**, *140*, 109523. [CrossRef]
6. Kohavi, R.; John, G.H. Wrappers for feature subset selection. *Artif. Intell.* **1997**, *97*, 273–324. [CrossRef]
7. Jiao, L.; Liu, Y.; Zou, B. Self-organizing dual clustering considering spatial analysis and hybrid distance measures. *Sci. China Earth Sci.* **2011**, *54*, 1268–1278. [CrossRef]
8. Chakraborty, B.; Chakraborty, G. Fuzzy Consistency Measure with Particle Swarm Optimization for Feature Selection. In Proceedings of the 2013 IEEE International Conference on Systems, Man and Cybernetics (SMC 2013), Manchester, UK, 13–16 October 2013; pp. 4311–4315.
9. Li, G.; Li, Y.; Tsai, C.-L. Quantile Correlations and Quantile Autoregressive Modeling. *J. Am. Stat. Assoc.* **2015**, *110*, 246–261. [CrossRef]
10. Pardo, L. New Developments in Statistical Information Theory Based on Entropy and Divergence Measures. *Entropy* **2019**, *21*, 391. [CrossRef]
11. Xue, B.; Zhang, M.; Browne, W.N.; Yao, X. A Survey on Evolutionary Computation Approaches to Feature Selection. *IEEE Trans. Evol. Comput.* **2015**, *20*, 606–626. [CrossRef]
12. Liu, Q.; Chen, C.; Zhang, Y.; Hu, Z. Feature selection for support vector machines with RBF kernel. *Artif. Intell. Rev.* **2011**, *36*, 99–115. [CrossRef]
13. Rong, M.; Gong, D.; Gao, X. Feature Selection and Its Use in Big Data: Challenges, Methods, and Trends. *IEEE Access* **2019**, *7*, 19709–19725. [CrossRef]
14. Abualigah, L.M.; Khader, A.T.; Al-Betar, M.A. Unsupervised feature selection technique based on genetic algorithm for improving the Text Clustering. In Proceedings of the 2016 7th International Conference on Computer Science and Information Technology (CSIT), Amman, Jordan, 13–14 July 2016; pp. 1–6.
15. Shamsinejadbabki, P.; Saraee, M. A new unsupervised feature selection method for text clustering based on genetic algorithms. *J. Intell. Inf. Syst.* **2011**, *38*, 669–684. [CrossRef]
16. Bennaceur, H.; Almutairy, M.; Alhussain, N. Genetic Algorithm Combined with the K-Means Algorithm: A Hybrid Technique for Unsupervised Feature Selection. *Intell. Autom. Soft Comput.* **2023**, *37*, 2687–2706. [CrossRef]
17. Zhang, Y.; Wang, S.; Ji, G. A Comprehensive Survey on Particle Swarm Optimization Algorithm and Its Applications. *Math. Probl. Eng.* **2015**, *2015*, 931256. [CrossRef]
18. Shami, T.M.; El-Saleh, A.A.; Alswaitti, M.; Al-Tashi, Q.; Summakieh, M.A.; Mirjalili, S. Particle Swarm Optimization: A Comprehensive Survey. *IEEE Access* **2022**, *10*, 10031–10061. [CrossRef]

19. Lalwani, S.; Sharma, H.; Satapathy, S.C.; Deep, K.; Bansal, J.C. A Survey on Parallel Particle Swarm Optimization Algorithms. *Arab. J. Sci. Eng.* **2019**, *44*, 2899–2923. [CrossRef]
20. Han, C.; Zhou, G.; Zhou, Y. Binary Symbiotic Organism Search Algorithm for Feature Selection and Analysis. *IEEE Access* **2019**, *7*, 166833–166859. [CrossRef]
21. Mohmmadzadeh, H.; Gharehchopogh, F.S. An efficient binary chaotic symbiotic organisms search algorithm approaches for feature selection problems. *J. Supercomput.* **2021**, *77*, 9102–9144. [CrossRef]
22. Cheng, M.-Y.; Prayogo, D. Symbiotic Organisms Search: A new metaheuristic optimization algorithm. *Comput. Struct.* **2014**, *139*, 98–112. [CrossRef]
23. Abdullahi, M.; Ngadi, A.; Dishing, S.I.; Abdulhamid, S.M.; Ahmad, B.I. An efficient symbiotic organisms search algorithm with chaotic optimization strategy for multi-objective task scheduling problems in cloud computing environment. *J. Netw. Comput. Appl.* **2019**, *133*, 60–74. [CrossRef]
24. Miao, F.; Zhou, Y.; Luo, Q. A modified symbiotic organisms search algorithm for unmanned combat aerial vehicle route planning problem. *J. Oper. Res. Soc.* **2018**, *70*, 21–52. [CrossRef]
25. Wu, H.; Zhou, Y.; Luo, Q. Hybrid symbiotic organisms search algorithm for solving 0–1 knapsack problem. *Int. J. Bio-Inspired Comput.* **2018**, *12*, 23–53. [CrossRef]
26. Baysal, Y.A.; Ketenci, S.; Altas, I.H.; Kayikcioglu, T. Multi-objective symbiotic organism search algorithm for optimal feature selection in brain computer interfaces. *Expert Syst. Appl.* **2020**, *165*, 113907. [CrossRef]
27. Gharehchopogh, F.S.; Shayanfar, H.; Gholizadeh, H. A comprehensive survey on symbiotic organisms search algorithms. *Artif. Intell. Rev.* **2019**, *53*, 2265–2312. [CrossRef]
28. Ganesh, N.; Shankar, R.; Čep, R.; Chakraborty, S.; Kalita, K. Efficient Feature Selection Using Weighted Superposition Attraction Optimization Algorithm. *Appl. Sci.* **2023**, *13*, 3223. [CrossRef]
29. Jaffel, Z.; Farah, M. A symbiotic organisms search algorithm for feature selection in satellite image classification. In Proceedings of the 2018 4th International Conference on Advanced Technologies for Signal and Image Processing (ATSIP), Sousse, Tunisia, 21–24 March 2018; pp. 1–5.
30. Cheng, M.-Y.; Cao, M.-T.; Herianto, J.G. Symbiotic organisms search-optimized deep learning technique for mapping construction cash flow considering complexity of project. *Chaos Solitons Fractals* **2020**, *138*, 109869. [CrossRef]
31. Mohammadzadeh, H.; Gharehchopogh, F.S. Feature Selection with Binary Symbiotic Organisms Search Algorithm for Email Spam Detection. *Int. J. Inf. Technol. Decis. Mak.* **2021**, *20*, 469–515. [CrossRef]
32. Cheng, M.-Y.; Kusoemo, D.; Gosno, R.A. Text mining-based construction site accident classification using hybrid supervised machine learning. *Autom. Constr.* **2020**, *118*, 103265. [CrossRef]
33. Al-Tashi, Q.; Abdulkadir, S.J.; Rais, H.M.; Mirjalili, S.; Alhussian, H. Approaches to Multi-Objective Feature Selection: A Systematic Literature Review. *IEEE Access* **2020**, *8*, 125076–125096. [CrossRef]
34. Abdollahzadeh, B.; Gharehchopogh, F.S. A multi-objective optimization algorithm for feature selection problems. *Eng. Comput.* **2021**, *38* (Suppl. S3), 1845–1863. [CrossRef]
35. Zhang, M.; Wang, J.-S.; Liu, Y.; Song, H.-M.; Hou, J.-N.; Wang, Y.-C.; Wang, M. Multi-objective optimization algorithm based on clustering guided binary equilibrium optimizer and NSGA-III to solve high-dimensional feature selection problem. *Inf. Sci.* **2023**, *648*, 119638. [CrossRef]
36. Al-Tashi, Q.; Abdulkadir, S.J.; Rais, H.M.; Mirjalili, S.; Alhussian, H.; Ragab, M.G.; Alqushaibi, A. Binary Multi-Objective Grey Wolf Optimizer for Feature Selection in Classification. *IEEE Access* **2020**, *8*, 106247–106263. [CrossRef]
37. Xue, B.; Fu, W.; Zhang, M. Differential evolution (DE) for multi-objective feature selection in classification. In Proceedings of the GECCO' 14: Genetic and Evolutionary Computation Conference, Dunedin, New Zealand, 15–18 December; pp. 83–84.
38. Vieira, S.M.; Sousa, J.M.C.; Runkler, T.A. Multi-criteria ant feature selection using fuzzy classifiers. In *Swarm Intelligence for Multi-objective Problems in Data Mining*; Springer: Berlin/Heidelberg, Germany, 2009; pp. 19–36. [CrossRef]
39. Xue, B.; Zhang, M.; Browne, W.N. Particle swarm optimisation for feature selection in classification: Novel initialisation and updating mechanisms. *Appl. Soft Comput.* **2014**, *18*, 261–276. [CrossRef]
40. Abdullahi, M.; Ngadi, A.; Dishing, S.I.; Abdulhamid, S.M.; Usman, M.J. A survey of symbiotic organisms search algorithms and applications. *Neural Comput. Appl.* **2019**, *32*, 547–566. [CrossRef]
41. Ezugwu, A.E.; Adewumi, A.O. Soft sets based symbiotic organisms search algorithm for resource discovery in cloud computing environment. *Future Gener. Comput. Syst.* **2017**, *76*, 33–50. [CrossRef]
42. Ezugwu, A.E.-S.; Adewumi, A.O. Discrete symbiotic organisms search algorithm for travelling salesman problem. *Expert Syst. Appl.* **2017**, *87*, 70–78. [CrossRef]
43. Ezugwu, A.E.-S.; Adewumi, A.O.; Frîncu, M.E. Simulated annealing based symbiotic organisms search optimization algorithm for traveling salesman problem. *Expert Syst. Appl.* **2017**, *77*, 189–210. [CrossRef]
44. Mohammadzadeh, H.; Gharehchopogh, F.S. A multi-agent system based for solving high-dimensional optimization problems: A case study on email spam detection. *Int. J. Commun. Syst.* **2020**, *34*. [CrossRef]
45. Arora, S.; Anand, P. Binary butterfly optimization approaches for feature selection. *Expert Syst. Appl.* **2018**, *116*, 147–160. [CrossRef]
46. Du, Z.-G.; Pan, J.-S.; Chu, S.-C.; Chiu, Y.-J. Improved Binary Symbiotic Organism Search Algorithm with Transfer Functions for Feature Selection. *IEEE Access* **2020**, *8*, 225730–225744. [CrossRef]

47. Miao, F.; Yao, L.; Zhao, X. Symbiotic organisms search algorithm using random walk and adaptive Cauchy mutation on the feature selection of sleep staging. *Expert Syst. Appl.* **2021**, *176*, 114887. [CrossRef]
48. Kimovski, D.; Ortega, J.; Ortiz, A.; Baños, R. Parallel alternatives for evolutionary multi-objective optimization in unsupervised feature selection. *Expert Syst. Appl.* **2015**, *42*, 4239–4252. [CrossRef]
49. Liao, T.; Kuo, R. Five discrete symbiotic organisms search algorithms for simultaneous optimization of feature subset and neighborhood size of KNN classification models. *Appl. Soft Comput.* **2018**, *64*, 581–595. [CrossRef]
50. Apolloni, J.; Leguizamón, G.; Alba, E. Two hybrid wrapper-filter feature selection algorithms applied to high-dimensional microarray experiments. *Appl. Soft Comput.* **2016**, *38*, 922–932. [CrossRef]
51. Zare-Noghabi, A.; Shabanzadeh, M.; Sangrody, H. Medium-Term Load Forecasting Using Support Vector Regression, Feature Selection, and Symbiotic Organism Search Optimization. In Proceedings of the 2019 IEEE Power & Energy Society General Meeting (PESGM), Atlanta, GA, USA, 4–8 August 2019; pp. 1–5.
52. Gana, N.N.; Abdulhamid, S.M.; Misra, S.; Garg, L.; Ayeni, F.; Azeta, A. Optimization of Support Vector Machine for Classification of Spyware Using Symbiotic Organism Search for Features Selection. In *Lecture Notes in Networks and Systems*; Springer: Cham, Switzerland, 2022. [CrossRef]
53. Zhou, Y.; Wu, H.; Luo, Q.; Abdel-Baset, M. Automatic data clustering using nature-inspired symbiotic organism search algorithm. *Knowl.-Based Syst.* **2019**, *163*, 546–557. [CrossRef]
54. Yang, C.-L.; Sutrisno, H. A clustering-based symbiotic organisms search algorithm for high-dimensional optimization problems. *Appl. Soft Comput.* **2020**, *97*, 106722. [CrossRef]
55. Zhang, B.; Sun, L.; Yuan, H.; Lv, J.; Ma, Z. An improved regularized extreme learning machine based on symbiotic organisms search. In Proceedings of the 2016 IEEE 11th Conference on Industrial Electronics and Applications (ICIEA), Hefei, China, 5–7 June 2016; pp. 1645–1648.
56. Ikotun, A.M.; Ezugwu, A.E. Boosting k-means clustering with symbiotic organisms search for automatic clustering problems. *PLoS ONE* **2022**, *17*, e0272861. [CrossRef] [PubMed]
57. Acharya, D.S.; Mishra, S.K. A multi-agent based symbiotic organisms search algorithm for tuning fractional order PID controller. *Measurement* **2020**, *155*, 107559. [CrossRef]
58. Rajah, V.; Ezugwu, A.E. Hybrid Symbiotic Organism Search algorithms for Automatic Data Clustering. In Proceedings of the 2020 Conference on Information Communications Technology and Society (ICTAS), Durban, South Africa, 11–12 March 2020; pp. 1–9.
59. Chakraborty, S.; Nama, S.; Saha, A.K. An improved symbiotic organisms search algorithm for higher dimensional optimization problems. *Knowl.-Based Syst.* **2021**, *236*, 107779. [CrossRef]
60. Sherin, B.M.; Supriya, M.H. SOS based selection and parameter optimization for underwater target classification. In Proceedings of the OCEANS 2016 MTS/IEEE Monterey, Monterey, CA, USA, 19–23 September 2016; pp. 1–4.
61. Bsoul, Q.; Salam, R.A.; Atwan, J.; Jawarneh, M. Arabic Text Clustering Methods and Suggested Solutions for Theme-Based Quran Clustering: Analysis of Literature. *J. Inf. Sci. Theory Pract.* **2021**, *9*, 15–34. [CrossRef]
62. Mehdi, S.; Smith, Z.; Herron, L.; Zou, Z.; Tiwary, P. Enhanced Sampling with Machine Learning. *Annu. Rev. Phys. Chem.* **2024**, *75*, 347–370. [CrossRef] [PubMed]
63. Larsen, B.; Aone, C. Fast and effective text mining using linear-time document clustering. In Proceedings of the KDD99: The First Annual International Conference on Knowledge Discovery in Data, San Diego, CA, USA, 15–18 August 1999; pp. 16–22.
64. Sanderson, M. Test Collection Based Evaluation of Information Retrieval Systems. *Found. Trends Inf. Retr.* **2010**, *4*, 247–375. [CrossRef]
65. Mohd, M.; Crestani, F.; Ruthven, I. Evaluation of an interactive topic detection and tracking interface. *J. Inf. Sci.* **2012**, *38*, 383–398. [CrossRef]
66. Zobeidi, S.; Naderan, M.; Alavi, S.E. Effective text classification using multi-level fuzzy neural network. In Proceedings of the 2017 5th Iranian Joint Congress on Fuzzy and Intelligent Systems (CFIS), Qazvin, Iran, 7–9 March 2017; pp. 91–96.
67. Lewis, D.D.; Yang, Y.; Rose, T.G.; Li, F. RCV1: A new benchmark collection for text categorization research. *J. Mach. Learn. Res.* **2004**, *5*, 361–397.

algorithms

MDPI

Article

Enhancing Indoor Positioning Accuracy with WLAN and WSN: A QPSO Hybrid Algorithm with Surface Tessellation

Edgar Scavino [1,*], Mohd Amiruddin Abd Rahman [1], Zahid Farid [2], Sadique Ahmad [3] and Muhammad Asim [3,*]

1 Department of Physics, Faculty of Science, Universiti Putra Malaysia (UPM), Serdang 43400, Malaysia; mohdamir@upm.edu.my
2 Department of Electrical Engineering, Abasyn University, Peshawar 25000, Pakistan; zahid.farid@abasyn.edu.pk
3 EIAS Data Science Lab, College of Computer and Information Sciences, Prince Sultan University, Riyadh 11586, Saudi Arabia; ahmad01.shah@ieee.org
* Correspondence: edgarscavino@upm.edu.my (E.S.); masim@psu.edu.sa (M.A.)

Abstract: In large indoor environments, accurate positioning and tracking of people and autonomous equipment have become essential requirements. The application of increasingly automated moving transportation units in large indoor spaces demands a precise knowledge of their positions, for both efficiency and safety reasons. Moreover, satellite-based Global Positioning System (GPS) signals are likely to be unusable in deep indoor spaces, and technologies like WiFi and Bluetooth are susceptible to signal noise and fading effects. For these reasons, a hybrid approach that employs at least two different signal typologies proved to be more effective, resilient, robust, and accurate in determining localization in indoor environments. This paper proposes an improved hybrid technique that implements fingerprinting-based indoor positioning using Received Signal Strength (RSS) information from available Wireless Local Area Network (WLAN) access points and Wireless Sensor Network (WSN) technology. Six signals were recorded on a regular grid of anchor points covering the research surface. For optimization purposes, appropriate raw signal weighing was applied in accordance with previous research on the same data. The novel approach in this work consisted of performing a virtual tessellation of the considered indoor surface with a regular set of tiles encompassing the whole area. The optimization process was focused on varying the size of the tiles as well as their relative position concerning the signal acquisition grid, with the goal of minimizing the average distance error based on tile identification accuracy. The optimization process was conducted using a standard Quantum Particle Swarm Optimization (QPSO), while the position error estimate for each tile configuration was performed using a 3-layer Multilayer Perceptron (MLP) neural network. These experimental results showed a 16% reduction in the positioning error when a suitable tile configuration was calculated in the optimization process. Our final achieved value of 0.611 m of location incertitude shows a sensible improvement compared to our previous results.

Keywords: quantum particle swarm optimization; indoor localization; fingerprinting; WiFi; wireless sensor network; multilayer perceptron; surface tessellation

1. Introduction

In recent years, the fast development of wireless technologies and their ubiquitous applications [1–3] have drawn a lot of attention for both outdoor and indoor localization tracking implementations [4–8]. As far as outdoor applications are concerned, the most known and popular technology is the satellite-based Global Position System (GPS), which offers uncountable services in the domains of tourism, navigation, military, etc. [9]. However, because of the satellite origin of the GPS signals, this technology loses performance in urban canyons, near tall walls, and in bad weather due to the weakness of the received signals, becoming completely useless in many indoor and underground environments. Consequently, GPS becomes an unreliable and ineffective method in scenarios where precise indoor localization is required [10].

In order to overcome these limitations, wireless technologies are being widely implemented for indoor positioning purposes. These techniques include Radio Frequency Identification (RFID) [11,12], and Wireless Local Area Networks (WLANs) [13–15], Bluetooth [16], Wireless Sensor Networks (WSNs), among others [17–19]. Other methods, based on the angle-of-arrival and the time-of-arrival, depend on measurements that require additional hardware and are therefore less common [20,21].

Thanks to the diffusion of smartphones and the availability of free WiFi in most indoor public spaces, the implementation of many of these communication technologies can be easily and inexpensively undertaken. As a consequence, location-based services are rapidly expanding with several benefits, mainly driven by the common availability of the appropriate hardware, both in terms of signal generation and reception.

The most common approach to infer the position of a receiver in a place where many wireless signals are available is based on the Received Signal Strength (RSS), which can be measured without the need to decode any communication feature. Localization is consequently determined by the pattern matching of the intensity of multiple WLAN sources [22]. Nevertheless, even the hybrid technology of a WiFi and WLAN combined approach for indoor positioning algorithms can lead to unavoidable incertitude, mainly driven by signal instability and fading, reflection on structures, the presence of people, and possibly autonomous trolleys [23,24]. In this work, the localization approach was based on the fingerprinting matching process from both WLAN and WSN signals as the basic scheme of positioning determination. The two technologies complemented each other in terms of spatial behavior, as WiFi signals penetrate walls, while WLAN presents a shorter range with a stronger gradient [25].

While a pattern-matching approach is classified as a deterministic algorithm [26], the spatial distributions of the signal intensities are strongly non-linear in the whole research space and thus require the implementation of methods such as neural networks and heuristic optimization techniques. Consequently, an Artificial Neural Network (ANN) approach guarantees robustness against noise and interference, which are among the major factors affecting the accuracy of indoor positioning systems.

In this work, we extended the concept of surface tessellation, in which we subdivided the research area into identical, discrete rectangular tiles. This approach was based on the idea that most of the time, a coarse location precision was sufficient to infer the actual position of a receiver and its carrier. In our previous work on research area tessellation [27], square tiles of 1 m side were considered as proof of concept. The published results, together with the improvements obtained by signal weighing and an appropriate ANN architecture, encouraged us to further investigate the tessellation technique for indoor positioning.

The core contributions of this paper are listed as follows:

1. To extend previously published results on the same dataset by joining our signal optimization results and the concept of tessellation of the search space.
2. To develop an optimization algorithm by investigating a suitable tessellation of the research space to reduce the overall error in the Multilayer Perceptron positioning algorithm.
3. To introduce a heuristic approach for the determination of regular space meshes in general positioning algorithms. With this fact in mind, a Quantum Particle Swarm Optimization (QPSO) algorithm was integrated into the positioning neural network application.

The remaining sections of the paper are organized as: Section 2 gives a brief description of the related work, including our previous analysis of the studied dataset. Section 3 presents in more detail the approach and techniques we used in this research. Here, particular attention is given to the QPSO process and to the improvement of the MLP application in space tessellation, which was used in our previously published works. The following part of this paper presents the results of this study, with references to the optimization of the MLP and the performance of our proposed QPSO algorithm. The achievements of our optimization process are presented and compared with results both from the literature and from the authors' previous research. Finally, the conclusions of this work are briefly discussed, and projects and ideas for future research are introduced.

2. Related Work

WiFi technology has been extensively researched for both communication and indoor positioning systems. However, relying solely on WiFi can result in poor performance due to signal fading and distortions caused by infrastructure and human presence. To address these issues, additional readily available sources of wireless signals, such as magnetic fields, inertial measurement units, and wireless sensor networks (WSN), have been incorporated.

Chen et al. [28] integrated multiple sensors, including an Inertial Measurement Unit (IMU), a magnetometer, and a WLAN RSSI (RSS Indicator). They proposed a Euclidean matching algorithm to estimate the location based on the WiFi signal. The Robust Extended Kalman Filter algorithm was used to estimate the location based on the IMU sensor. Additionally, an enhanced dynamic time-warping algorithm is used to match the magnetic fingerprint location based on magnetometer measurement. Although multiple sensor systems improved the accuracy of the positioning system, the computation required user mobile processing power to obtain the location information, which might be unfeasible for real-time applications.

Luo et al. [29] proposed a joint WSN–WiFi approach for improving indoor positioning estimation. The work considered the communication between sensor nodes as well as between the sensor nodes and the receiver to help with user positioning. The detection algorithm considered a grid-based search, with the computation cost increasing as the search location was expanded due to the increment in the number of grids to be considered. Khan et al. [27] also proposed a WSN–WiFi technique based on a grid search that considered the floor tile spacing. This technique estimates the location of the user to be within the respective boundaries of the specific tile's location based on an appropriately trained neural network.

Guo et al. [30] included WiFi and cellular telephony network RSS signals in a hybrid approach for an indoor positioning algorithm. The fingerprinting database was analyzed by Principal Component Analysis (PCA) with interpolations of the raw data. Subsequently, data were processed by an ANN separately as cellular signals were first used to provide coarse training, which was then improved using the WiFi data. This hybrid approach improved the positioning precision; however, the process was very demanding in terms of computing time and unsuitable for real-time applications.

The synergy between multiple smartphone sensors was investigated by Gang and Pyun [31], who joined the signals of Bluetooth, magnetic field sensors, IMU, and a camera together. Their accomplishments demonstrated the effectiveness of a hybrid approach, in which combined heterogeneous sources of signals provided better indoor positioning results. However, their method required a permanent connection to a central server, and they chose not to use WiFi signals, which were deemed to be variable and unreliable in RSS fingerprinting algorithms.

This work implements a consolidated hybrid approach as a positioning system by means of combined RSS signals from WiFi and WSNs in order to improve localization accuracy in indoor spaces. The utilization of WiFi–WSN signals does not require any additional computation in end-user devices and can be used for real-time applications. In this work, we apply the results of our previous research on relative signal weighing and MLP architecture optimization, which led to an overall significant reduction of the positioning error on the same dataset [32]. We also utilize a similar algorithm based on standard QPSO, which we integrate with the MLP neural network in order to improve the accuracy of the localization system. The detailed method, which aims to find values of appropriate tile sizes and tessellation corners, is introduced in the next section.

3. Research Methodology

Due to their widespread deployment and access point availability, WLAN and WSN-based technologies are gaining interest in the domain of indoor localization and positioning services. RSS data from WLAN and WSN sources are becoming increasingly important for refining and improving the accuracy in indoor environments [13,17], particularly when GPS signals become unreliable or undetectable. In this work, our goal consists of the implementation of a refinement of previous research, again by using a hybrid indoor positioning system in which both signals from available WLAN access points and WSN-based technologies are considered.

In particular, the possibility of further refinement and optimization is based on two previous distinct works, in which we separately considered the possibility of weighting the WLAN and WSN source signals and the tessellation of the research space into a regular grid of square tiles. A preliminary inspection of the collected data confirmed the lack of linearity of signal intensities with respect to the relative positions inside the research area. This aspect reinforced the need to process the data by means of a machine learning approach, in which we included the approach of weighting the available signals from both WLAN and WSN with the aim of improving indoor positioning accuracy.

The concept of tessellation of the research space is based on the fact that, for many applications, the precise localization of a signal receiver and its end-user is not required. Furthermore, any positioning algorithm presented and discussed in the literature carries an incertitude of the order of magnitude of 1 m. As a result, for most indoor positioning applications, finite-sized tiling would not become a limiting factor in precision for localization purposes, especially in large indoor environments. For this reason, this work focuses on the optimization of a regular tessellation of the research space, seeking optimal sizes of rectangular adjacent tiles and the collocation of the corner of the resulting grid.

3.1. Signal Acquisition

In order to address the subject of an accurate indoor positioning system, raw data collection, and preprocessing are significant steps for location fingerprinting. This study investigates a collection of data that were measured as RSSI values based on an IEEE 802.11b/g wireless card. RSSI data were acquired at the main basement level of the Faculty of Engineering and Built Environment at the Universiti Kebangsaan Malaysia (UKM) [33]. In January 2014, data were collected on a single day without interruption. While the research area was regular in shape and did not present any obstacles, the boundaries were quite heterogeneous, with research laboratories, furnished offices, lecture halls, structural pillars, and people occasionally crossing the place. The research area and the schedule of the data acquisition were purposely chosen in order to reproduce the normal conditions of a typical working day. This way, disturbances in the signals were taken into account as they were received and measured.

Figure 1 shows a sketch of the research area and its irregular boundaries. A regular grid including 96 anchor points encompasses the unobstructed part, in which WLAN and WSN signals were recorded regardless of people and materials transiting through the experimental area. For experimental RSS data collection, we selected a regular grid of 96 anchor points for the creation of the fingerprinting database, in an array of 6×16 points, as a consequence of the chosen horizontal and vertical separation between anchor points, respectively, 1 m and 1.5 m in the directions parallel to the main walls. The height of each anchor point was strictly kept at the value of 1 m in order to avoid variations in the locally measured signal intensities and to remove the height as a possible sensitive variable [34].

A Lenovo G580 laptop with a core i5 processor and Windows 7 operating system was used as a measurement setup. For the collection of RSS data samples, an Atheros AR9285 onboard wireless adaptor based on 802.11n was added. An "inSSIDer" open WiFi Scanner software (version 4.2) was used, to which some adaptations were executed in order to create the log files of the measured RSS signals with respect to the visible MAC address and the time stamp.

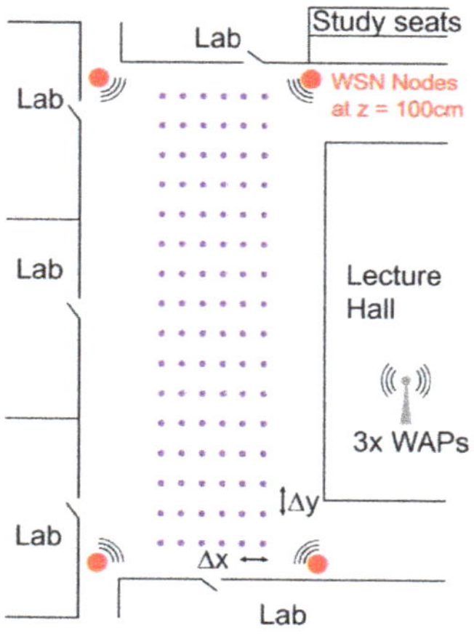

Figure 1. A sketch of the research area, with depictions of irregular boundaries, signal sources, and the grid of anchor points where signal intensities were recorded.

The signal intensities in dB from three available WiFi and four installed WSN signal sources were recorded simultaneously 300 times for one minute, i.e., every 0.2 s, in each of the 96 anchor points. Only three signals from the WSN were considered for the database, as one of the access points was faulty and provided random, unreliable values. The average of each set of 300 values was used to build the signal intensity database. The stability of all six available signals was checked through the standard deviation of each set, which was satisfactorily within 5% of the average value for all anchor points. The acquired data were organized and saved in a matrix of 96×8 elements. The WiFi data were stored in columns 1–3; the WSN data were placed in columns 4–6; and columns 7 and 8 contained the spatial coordinates of the 96 anchor points. In order to utilize the appropriate signal weighting, which provided a significant reduction in the positioning error in our previous work [32], it was strictly necessary to identify and keep the same signal sequence in all matrix columns. At the same time, the meaning of this research is fundamentally dependent on the constant positions of all access points and of any other signal source.

3.2. Data Analysis

The procedural scheme of the proposed optimization algorithm is based on a number of fundamental sections. First, data from hybrid sources were acquired and stored using fingerprinting and RSS principles. Data were kept as they were recorded, without any particular preprocessing, any further analysis, or suggestion of a limited database.

The second step consisted of creating a mathematical model for both the distribution of the signal data and a mainframe for the tessellation of the research space. As the behavior of the measured signal intensities along the 2D space of the experimental area proved to be inherently non-linear, a Multilayer Perceptron (MLP) neural network was developed and trained thanks to its versatility in modeling inherently non-linear data. In consideration of the supervised training features of any MLP, the target was identified as the preprocessed position of the 96 anchor points. Given a tessellation configuration with a predetermined tiling position and corner point, each anchor point would be assigned to the tile to which it belonged geometrically. The MLP target would then become a set of discrete, integer pairs of coordinates.

Finally, an optimization algorithm was developed to address the main subject of this research, i.e., to investigate the possibility of reducing the positioning error through the MLP by means of an appropriate choice of the tile size and collocation of the tessellation corner point. Taking into account the number of anchor points and the possible variability of the tessellation geometry, a heuristic approach to the optimization process was selected.

A QPSO algorithm was chosen due to its features of rapid convergence to a global optimal solution and its relative ease of programming when compared to other available optimization algorithms [35].

3.3. QPSO Process

The role of the QPSO in data analysis in this research consists of searching for an appropriate set of four variables that define the tessellation properties of the search space, namely the tile width and height, and the horizontal and vertical positions with respect to the bottom-left corner point.

In most "swarm-based" optimization processes, a population of candidate solutions is first created within a multidimensional search space. Every candidate solution, also named "individual" or "particle", is associated with a cost value, which is evaluated according to the particular optimization task. The resulting individual in this QPSO process thus becomes a four-dimensional vector of homogeneous geometrical features. At every iteration, each individual evolves its position in the search space with a law of motion linked to two attractors, namely its own best value and the overall best value that is calculated among the whole population and updated at every loop.

The QPSO algorithm, first introduced in 2004 [36,37], is inspired by the Quantum Mechanics concept that every particle is associated with a wave and the fact that quantum waves cover all the search space. The position of each individual is updated in the spirit of Quantum Mechanics and relies heavily on the generation of random numbers. The position update for each individual is governed by two attractors, the average value of all N individuals' best positions:

$$\boldsymbol{m} = \frac{1}{N}\sum_{i=1}^{N} \boldsymbol{b}_i \tag{1}$$

and a linear combination with random coefficients c_1 and c_2 of the individual and global best:

$$\boldsymbol{p}_i = (c_1\boldsymbol{b}_i + c_2\boldsymbol{g})/(c_1 + c_2) \tag{2}$$

The position update for iteration $(n+1)$ is then provided by the following expression, which identifies the maximum likelihood coordinated where to find the individual:

$$\boldsymbol{x}_i(n+1) = \boldsymbol{p}_i + \beta(n)\cdot\text{sign}(c_3 - 0.5)\cdot\ln(c_4)|\boldsymbol{m} - \boldsymbol{x}_i(n)| \tag{3}$$

where c_3 and c_4 are uniformly distributed random coefficients and $\beta(n)$ is a uniformly decreasing parameter that traditionally starts at $\beta(1) = 1$ and reaches $\beta = 0.1$ after a preset number of iterations, at which the algorithm ends.

Considering that the QPSO procedure involves individuals and positions evaluated by a linear combination of vectors, its applicability is naturally taken into consideration, as in this work the search space is a subset of the multidimensional set $\mathbb{R}^4$. The cost function $f(x_i)$ was evaluated for each individual $x_i = (w_i, h_i, u_i, v_i)$ using an MLP neural network, which was designed to provide the most accurate positioning achievable through appropriate network training.

3.3.1. MLP–ADAM-Based Localization

An Artificial Neural Network approach [38] is justified by the observation that the variations in each of the six meaningful signals along the research area are strongly non-linear and unpredictable due to their complexity. A commonly used ANN architecture is the simple, fully connected, feedforward MLP network, which utilizes non-linear filtering functions. MLP training is usually performed by the backpropagation technique, which is a momentum-improved gradient-descent reduction process of the regression error.

The ultimate goal of any MLP training consists of reducing the error computed between the network output and a preset target in a supervised learning procedure. The available input data is usually split into two subsets, namely the training data and the test data. The network is trained by altering the weight matrixes that characterize the

MLP architecture by using the {input vector—expected target} pairs of the training data alone. The pairs {test output—expected target} are used as a form of quality control for the training in terms of the verification of the network generalization vs. memorization capabilities.

In this work, three WiFi signals and three WSN signals were collected from 96 anchor points in a regular array of positions in the research area. All the experimental data were included in the training set, while the test set was created as a list of 20 points randomly selected within the experimental area. The test signals were generated by performing standard bilinear interpolation on the actual measured values in the vertices of the rectangle containing each of the test points. Figure 2 (left) shows a typical layout of the training and testing points within the experimental area.

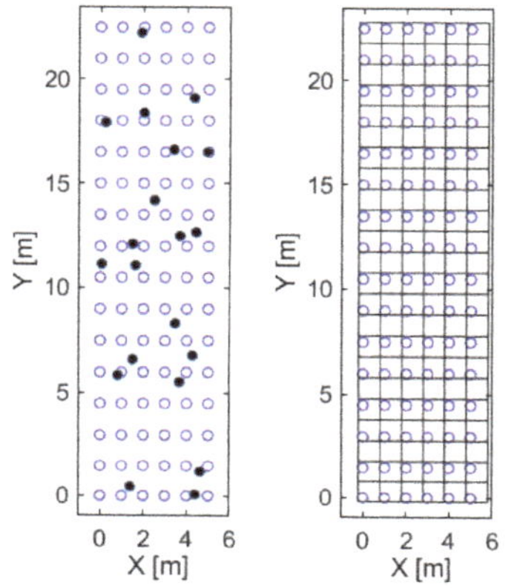

Figure 2. **(left)** The layout of the anchor points (blue circles) and test points in random positions (black dots) within the experimental area. **(right)** A tessellation of the research area is shown, with respect to the anchor points.

The preprocessing of the input data before feeding them into the MLP involved two steps. First, a relative weighting of the signal intensities was applied based on our previously published results on the same dataset [32]. Specific weights were assigned to each of the six signals. In the second step, the average of each individual signal intensity was subtracted from the weighted values to extract and retain their meaningful spatial variations using the following equation:

$$S'_j = S_j - \frac{1}{N}\sum_{i=1}^{N} S_{ji} \tag{4}$$

for $j = 1$–6 and $N = 116$, as the 20 test data were included in Equation (4) for consistency in the training process.

The preprocessing of the target data required the transformation of the available spatial coordinates into a discrete representation, following the concept of surface tessellation. The tile sizes and the offset position of the tessellation are the goals of the optimization process of this research. Thus, the transformation from spatial to discrete coordinates is fundamentally dependent on the individual candidate solutions, which are managed by the QPSO algorithm. Nonetheless, the preprocessing of the target followed a standard procedure in which each anchor point is univocally assigned to the tile it lies on geometrically. First, the integer number of tiles N_x and N_y are calculated, respectively, for x and y directions:

$$\begin{cases} N_x = \left\lceil \frac{x_{max}-u}{w} \right\rceil \\ N_y = \left\lceil \frac{y_{max}-v}{h} \right\rceil \end{cases} \tag{5}$$

where x_{max} and y_{max} are the maximum positional coordinates or any train or test data. The partial square bracket symbol $\lceil \cdot \rceil$ indicates the rounding up to the next integer, i.e., the number of tiles. The values u and v—offset coordinates where the tessellation starts; w, h—the width and height of the tiles are all passed to the MLP as parameters by the QPSO algorithm. Consequently, each anchor point or test point of coordinates (x_i, y_i) is associated with the discrete 'tile' values:

$$\begin{cases} n_x = \left\lceil \frac{x - x_{off}}{u} \right\rceil \\ n_y = \left\lceil \frac{y - y_{off}}{v} \right\rceil \end{cases} \tag{6}$$

Figure 2 (right) shows an example of virtual tessellation of the research space, in which square tiles with a side of 1 m have been superposed on the array of experimental points, with an offset of 20 cm in both directions, resulting in an array of 6 × 22 tiles.

The last step of the target preprocessing is required by the fact that the output of the standard MLP network lies in the interval [0, 1], thus a linear normalization of the set of values n_x and n_y must be performed according to:

$$\begin{cases} n'_x = 0.1 + 0.8\frac{n_x - 1}{N_x - 1} \\ n'_y = 0.1 + 0.8\frac{n_y - 1}{N_y - 1} \end{cases} \tag{7}$$

This rescaling guarantees that the set of normalized targets lies in the range [0.1, 0.9], thus optimizing the performance of the training process of the MLP in terms of computing speed and target separation.

The training data are used in the standard feed forward-backpropagation learning process, which aims to find appropriate weight matrices W_k that best model the neural network on the spatial distribution of the experimental data. The output of the testing data is used for the training termination criterion, whereas the testing output, calculated with the same weight matrices, is not used for training. The testing error is obtained and stored, and the MLP training process is deemed complete when the relative positioning error reaches its minimum value. The first step of the feedforward phase consists of calculating the first hidden layer h_1 using the whole input data X organized in a matrix of 96 rows—the signals in each point—and six columns—the six signals in each point—thus using the batch training process:

$$\boldsymbol{h}_1 = f(\boldsymbol{X}\boldsymbol{W}_1) \tag{8}$$

In this expression, $f(\cdot)$ is the filter or activation function that expresses the non-linearity feature of an MLP network. In this work, we used the standard sigmoid function $f(x) = \frac{1}{1+\exp(-x)}$.

In our previous research using the same experimental data [32], we determined that the best results were achieved with three hidden layers, thus here we keep the same architecture for this work. The second and third hidden layers as well as the output are then calculated according to Equation (9):

$$\begin{array}{r} \boldsymbol{h}_{k+1} = f(\boldsymbol{h}_k \boldsymbol{W}_{k+1}) \\ \boldsymbol{out} = f(\boldsymbol{h}_3 \boldsymbol{W}_4) \end{array} \tag{9}$$

The same process is performed separately for the test data. The MLP errors for both training and testing are calculated as $err = \frac{1}{2}\sum\sum(\boldsymbol{out} - \boldsymbol{tar})^2$ where the sum extends to all elements of the output and target matrices. The standard MLP training is based on the gradient descent process in which the partial derivatives $\Delta \boldsymbol{W}_k = \partial err / \partial \boldsymbol{W}_k$ of the weight matrixes are calculated and the weights W_k are updated according to the expression:

$$\boldsymbol{W}_k^{(n+1)} = \boldsymbol{W}_k^{(n)} + \lambda \Delta \boldsymbol{W}_k^{(n)} + \eta \Delta \boldsymbol{W}_k^{(n-1)} \tag{10}$$

where λ is the learning rate and μ is the momentum, a term that is included to speed up the training process by taking into account the correction applied at the previous iteration.

In this work, we employed a refinement of the backpropagation algorithm, i.e., the novel adaptative momentum technique, or 'adam' [39], which is a proven method for expediting the learning process of the MLP. In this process, at every loop the update of each element in the weight matrices $\mathbf{W}_k$ is proportional to the partial derivative $\partial err/\partial W_{kij}$ through a variable learning rate factor, which increases the training speed in flat regions of the MLP error. Two constants, β_1 and β_2, are introduced, as well as a 'first' and a 'second' set of momentum matrices, $\boldsymbol{\eta}$ and $\boldsymbol{\mu}$, respectively. The equations for the update of the first and second momenta are:

$$\begin{aligned} \boldsymbol{\eta}_k^{(n+1)} &= \beta_1 \boldsymbol{\eta}_k^{(n)} + (1-\beta_1)\Delta \mathbf{W}_k \\ \boldsymbol{\mu}_k^{(n+1)} &= \beta_2 \boldsymbol{\mu}_k^{(n)} + (1-\beta_2)(\Delta \mathbf{W}_k)^2 \end{aligned} \tag{11}$$

and the weights are updated according to the expression:

$$\mathbf{W}_k^{(n+1)} = \mathbf{W}_k^{(n)} - \lambda \frac{(1-\beta_2^n)\boldsymbol{\eta}_k^{(n+1)}}{(1-\beta_1^n)\sqrt{\boldsymbol{\mu}_k^{(n+1)}}} \tag{12}$$

3.3.2. Positioning Algorithm

In this research work, the positioning algorithm was developed within the frame of a search space quantization, in which a regular tessellation is generated. Training and testing points are assumed to belong to a rectangular tile whose spatial coordinates are numbered with integer values.

The goal of the optimization process consists of finding optimal horizontal and vertical dimensions for the tiles, as well as an optimal starting point for the surface tessellation. For any given geometrical configuration, each experimental point belongs to a specific tile. The MLP was designed with the task of reducing—by Adam-driven descent—the error:

$$err = \frac{1}{2}\sum\sum(\mathbf{out} - \mathbf{tar})^2 \tag{13}$$

as per the algorithm described in Section 3.3. The process was terminated after a predetermined number of iterations when the minimum value of the real space positioning error for the test data was supposedly reached. For this reason, several steps had to be performed at regular intervals during the MLP iterations. First, the output out_{test} of the test data was calculated and rescaled by inverting the expressions in Equation (7), and the result was rounded to the nearest integer in order to identify the output tile the test data was assigned to. Then, the horizontal and vertical tile mismatches were evaluated for each test point, with the absolute difference between the output tile numbers $[n_{x,out},\ n_{y,out}]$ and the target values $[n_{x,tar},\ n_{y,tar}]$, which were calculated in Equation (6):

$$\begin{cases} \Delta \boldsymbol{n}_x = |\boldsymbol{n}_{x,out} - \boldsymbol{n}_{x,tar}| \\ \Delta \boldsymbol{n}_y = |\boldsymbol{n}_{y,out} - \boldsymbol{n}_{y,tar}| \end{cases} \tag{14}$$

Finally, the real horizontal and vertical mismatches were calculated, and the error in the test data was evaluated as the average of the distance errors in Euclidean metrics in the set of all test points:

$$err_{real} = \langle \sqrt{(w\Delta n_x)^2 + (h\Delta n_y)^2} \rangle \tag{15}$$

In order to perform a meaningful comparison between tiles of different dimensions during the optimization process, an amount was added on the basis that a point randomly located in a rectangular tile and relocated to the center carries an intrinsic error given by:

$$err_{[h,w]} = \iint \sqrt{x^2+y^2}dxdy = \frac{1}{6}\left[\sqrt{h^2+w^2} + \frac{h^2}{2w}ln\frac{\sqrt{h^2+w^2}+w}{h} + \frac{w^2}{2h}ln\frac{\sqrt{h^2+w^2}+h}{w}\right] \quad (16)$$

where the integral is calculated in a rectangle of dimensions $[h,\ w]$.

4. Analysis and Experimental Results

This section is devoted to providing detailed analysis and results obtained using the proposed algorithm.

4.1. Experimental Setup

In order to achieve the stated goal, a standard QPSO algorithm was designed with the parameters shown in Table 1, which are used as standard values capable of speeding up the overall QPSO procedure. The search range for the tile dimensions was limited to the interval (0.3, 1.2) m for both horizontal and vertical dimensions, while the offset was limited to the actual tile size to avoid placing tiles without experimental data in the lower-left corner of the research area.

Table 1. QPSO design and architecture parameters for tile size optimization.

Design Parameters		Architecture Parameters	
Number of individuals	20	c_1	1
Number of iterations	100	c_2	1.5
Number of cost calculations	2000	Initial β_0	1
Search range (tile size)	(0.3, 1.2)	Final β_0	0.1

The algorithm for the proposed technique was developed in Matlab and ran on a Windows-based 2.6 GHz Core i7 desktop. The QPSO algorithm, like any other heuristic optimization process, is supposed to provide better search results using larger populations and a greater number of generations. However, for this work, each cost calculation took an average of 50 s; thus, the values reported in Table 1 were accepted as a good compromise between computing time and the quality of the QPSO capabilities.

The outcome of any MLP is usually heavily dependent on the initial random set of the training weight matrixes. Consequently, a large amount of MLP training runs are required to identify suitable initial weight choices and achieve meaningful results.

A number of validation runs were performed to investigate the effect of the random positioning of the 20 test points inside the research area. In agreement with the investigation presented in our previous research [32], no meaningful difference was observed in the positioning error for constant test points versus random ones. Both in these validation runs and in the actual QPSO algorithm, the MLP was run 1000 times in order to achieve the best performances, i.e., to identify a suitable set of weights for which the positioning error of the test population was minimal. For each training run, the profiles of both the training and testing positioning errors versus the learning iteration were recorded, and the absolute minimum of the testing error was identified, as shown in Figure 3.

It was observed during the validation runs that the absolute minimum of the test error was most probably achieved shortly after 4000 iterations. Consequently, it was decided to stop the MLP learning run after 5000 cycles of feedforward-backpropagation as a good trade-off between accuracy and total computing time. The cost value for each QPSO individual was evaluated as the average of the best 50 of 1000 MLP runs, which would correspond to fortunate initial random choices of the weights.

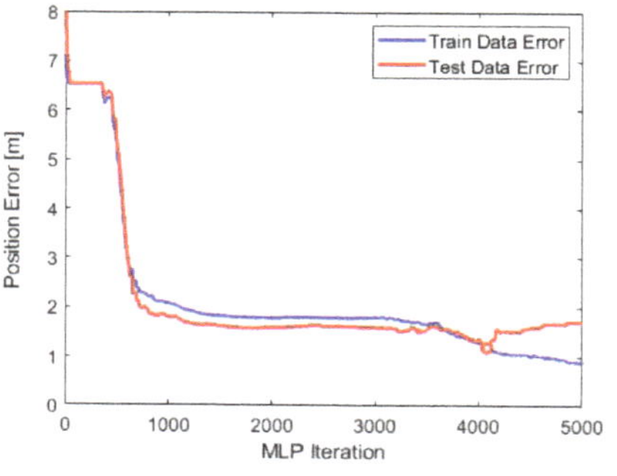

Figure 3. The profiles of the position error relative to the training and testing data in a typical MLP run. The minimum value in the test profile is highlighted.

4.2. Results

Following the definition of the optimization procedure using a QPSO algorithm and the technique for calculating the localization error using a standard MLP with an adaptative backpropagation process, the next step in this research was to select the most appropriate values for both the QPSO and MLP algorithms. Several parameters can affect the speed of both processes and strongly alter the total computing time. For the QPSO part, it was decided to choose the standard parameters found in the literature, namely those reported in Table 1. The QPSO procedure was repeated for a total of 100 generations, each producing a new population of 20 vectors of 4 components, recording the size of the tiles as well as the offset position with respect to the bottom left experimental point.

Due to the QPSO's nature, the global best individual, i.e., the vector of signal weights producing the lowest positioning error, was recorded at every stage of the algorithm. Each vector was first tested under the preset conditions, which limited the tile size to the interval (0.3, 1.2) m for both horizontal and vertical dimensions and the offset to the tile size. Each non-compliant configuration was discarded and the generation of a new individual was repeated until the preset population size was reached. The positioning errors of the new QPSO generation of individuals were then evaluated by repeating 1000 times the MLP procedure for each individual and calculating the average of the 50 best test positioning errors.

Multilayer Perceptron neural networks are subject to several parameters that greatly affect the training accuracy and the computing time. For this research, it was decided to refer to the investigation of the validation runs in our previously published research on the same dataset [32] and to keep the same MLP architecture. In particular, three hidden layers of 18 hidden units each were retained, as well as the standard sigmoid activation function. The values of the learning rates and momentum coefficients for the Adam technique were determined by visual inspection during a small number of dry runs, as well as by measuring the computing time of individual cost calculations. Moreover, the intensities of the input signals were weighted, in agreement with our previous results [32]. The set of parameters for the MLP process are reported in Table 2. Because of the sheer amount of QPSO cost evaluations and the large number of MLP runs for each individual, the overall computing time reached about 28 h despite the fact that each MLP training process of 5000 iterations took only 50 ms to complete.

The output of the MLP process was designed to return the integer values of the tile positions in the regular array defined by the vectors $x_i = (w_i, h_i, u_i, v_i)$, both for the train and the test data points. Equations (14)–(16) were then applied to calculate the actual position error for both data, while only the test position error was retained as the result of this research.

Table 2. MLP design and operational parameters.

Architecture Parameters		Training Parameters	
Hidden layers	3	Learning rate	$\eta = 0.001$
Hidden units	18 per layer	Momenta Adam values	$\beta_1 = 0.9$, $\beta_2 = 0.999$
Activation function	Logistic sigmoid	Runs per individual	1000
Training points	96	Training iterations	5000
Test points	20	Computing time	~28 h

A meaningful comparison of configurations with different tile sizes cannot be done on the basis of the number of tiles mismatch in either direction alone, as larger tiles might be correctly identified despite a relevant actual position error. For this reason, the value of the intrinsic error was taken into account as the average distance between a generic point and the center of the tile, according to Equation (16). Thus, the actual positioning error err_{pos}, i.e., the QPSO cost value for an individual becomes:

$$err_{pos} = err_{real} + err_{[h,w]} \tag{17}$$

in which the two parts are defined by Equations (15) and (16).

The value of the resulting positioning error as a function of the QPSO iteration number is shown in Figure 4 (right). It can be observed that the improvements in the positioning error take place in stages, in agreement with the discrete exploration technique of the QPSO. No error improvement was achieved beyond QPSO iteration #56, which provided a positioning error m measured on test data alone. This value shows an improvement of ~16% with respect to our previous best result, in which signal weighting was first introduced for the analysis of the same dataset.

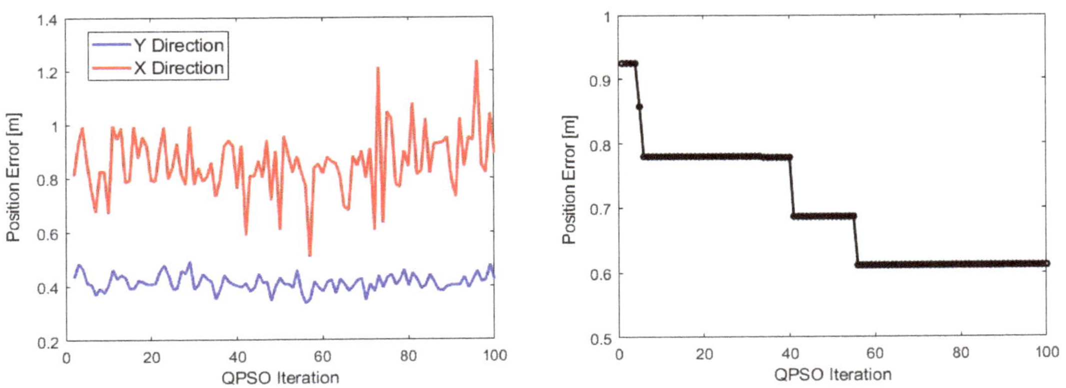

Figure 4. (left) The values of the horizontal and vertical contributions to the total positioning error, as calculated at each QPSO iteration. **(right)** The overall performance of the QPSO algorithm in terms of the positioning error reduction. A final value of 0.611 m was achieved.

The final geometrical configuration that led to this positioning error provided a tile size (w, h) = (0.505, 0.587) m and an offset (u, v) = (0.059, 0.411) m. These values correspond to a tessellation of 11 tiles horizontally and 40 vertically. The tile size provides an intrinsic error of 0.209 m as per Equation (16), thus leading to a tile error of 0.402 m. In consideration of the tile size and the procedure of calculating the positioning error, it can be concluded that, on average, each test data carries a mismatch of a single tile.

In our previously published work on the same dataset using fixed surface tessellation [27], it was observed that the positioning error was mainly caused by tile mismatches in the horizontal direction. For this reason, in this research, further attention was given to the separate contributions to the total positioning error by the horizontal and vertical tile mismatches. At every QPSO iteration, the position error in each direction was recorded for the best individual out of the population of 20. Again, it can be observed that in the horizontal direction, the partial position error is larger and more widely distributed compared to the values recorded in the vertical direction. Figure 4 (left) shows that in the vertical direction, the error does not vary much around 40 cm, while in the horizontal direction, values are spread in the range of 0.6–1.2 m. Nevertheless, the particularly low value recorded at QPSO iteration #56 generated the final global best for this research.

A comparison of the results obtained in this research with a selection of relevant previously published location average errors is presented in Table 3. Our previous work, based on the same raw data, is included together with other benchmark results available in the literature, showing a remarkable reduction in the positioning incertitude.

Table 3. Comparison of our obtained results with previously published research.

Reference	Chosen Approach and Implementation	Research Area and Anchor Points (AP)	Error
M. Ficco et al. [40]	Source selection, WiFi + Bluetooth	20 m × 21 m, 26 AP	>4 m
L. Jiang [26]	K-NN modeling, WiFi + RSS	15 m × 40 m, 100 AP	1.70 m
Z. Xiong et al. [41]	Particle filter, WiFi + RFID	25 m × 12 m, 80 AP	1.60 m
J. Chen et al. [28]	Optimization, IMU + WLAN	90 m path, 113 AP	1.48 m
Z. Farid et al. (same dataset) [33]	MLP modeling, WiFi + WSN	6 m × 24 m, 96 AP	1.22 m
I. U. Khan et al. (same dataset) [27]	Tessellation, WiFi + WSN	6 m × 24 m, 96 AP	1.01 m
E. Scavino et al. (same dataset) [32]	Optimization, WiFi + WSN	6 m × 24 m, 96 AP	0.725 m
This Research (same dataset)	Variable tessellation, WiFi + WSN	6 m × 24 m, 96 AP	0.611 m

The proposed method of data analysis does not present any particular challenges, apart from the extensive computing time. Nevertheless, the actual application of this technology would require some caveats about the stability and availability of the sources of the signals. Possible drifts and abrupt variations would render this method less accurate, thus requiring particular attention to the hardware of the whole installation. Further work beyond this method should possibly focus on the normalization of the input signals and the analysis of their spatial distributions.

5. Conclusions

This paper illustrates a QPSO algorithm for an indoor location system based on hybrid raw RSS signals, namely those from WiFi and WSN sources. The proposed system considered the possibility of reducing the positioning error by performing a regular tessellation of the research area in which the raw signals were acquired. Each QPSO individual was identified as a vector of four values containing the tile size and offset, which unambiguously determined a tessellation geometry. A Multilayer Perceptron network was developed as a tool to heuristically predict the positioning error for each QPSO individual. The architecture of the MLP, as well as the initial raw signal weighting, were kept in agreement with our previous research, thus allowing a faster network setup and positioning error reduction. The final results of the optimization process allowed us to identify an optimal geometrical configuration of the regular surface tessellation in which the positioning error showed a marked reduction compared to our previous research using the same set of data. We achieved a final localization accuracy of 0.611 m, representing an improvement of at least 16% compared to our previous best results using the same dataset. This value also marks an improvement of at least 50% over our initial published analysis of these data.

In addition to determining the overall positioning accuracy, a raw analysis of the spatial source of error was performed in order to identify the contribution along the horizontal and vertical directions. For every QPSO generation, in the horizontal direction, the error component was larger by roughly a factor of two compared to the vertical component, thus providing a large amount of the positioning error.

The goal of any indoor positioning algorithm consists of reducing the localization error in real-life scenarios in which randomly set wireless receivers are capable of precisely inferring their spatial coordinates. To this date, we have analyzed hybrid approaches with two sources, namely WiFi and WSN, and several algorithms, i.e., signal weighing with smooth research space and variable tessellation in this work. These results encourage further investigation, and in future work, all the developed approaches may be joined together, as well as including further techniques of spatial data analysis or an improved MLP training process that takes into account the different directional error contributions.

Author Contributions: Conceptualization, E.S., M.A.A.R., Z.F., S.A., and M.A.; Data curation, M.A.A.R., Z.F., and M.A.; Formal analysis, E.S. and Z.F.; Funding acquisition, S.A. and M.A.; Investigation, M.A.A.R., S.A., and M.A.; Methodology, E.S. and M.A.A.R.; Resources, M.A.A.R., Z.F., and M.A.; Software, E.S.; Supervision, M.A.A.R.; Validation, M.A.A.R., Z.F., S.A., and M.A.; Visualization, E.S. and M.A.; Writing—original draft, E.S.; Writing—review and editing, M.A.A.R., Z.F., S.A., and M.A. All authors have read and agreed to the published version of the manuscript.

Funding: This research was supported by EIAS Data Science & Blockchain Lab, Prince Sultan University. The authors would also like to thank Prince Sultan University for paying the APC for this article.

Data Availability Statement: Data can be shared upon reasonable request from the corresponding author.

Acknowledgments: The authors would like to thank Prince Sultan University for their support. In addition, the authors also wish to thank Rosdiadee Nordin and Mahamod Ismail of UKM for helping with the organization of data collection in public areas of the Faculty of Engineering.

Conflicts of Interest: The authors declare no conflicts of interest.

References

1. Wang, H.; Ning, H.; Lin, Y.; Wang, W.; Dhelim, S.; Farha, F.; Ding, J.; Daneshmand, M. A Survey on the Metaverse: The State-of-the-Art, Technologies, Applications, and Challenges. *IEEE Internet Things J.* **2023**, *10*, 14671–14688. [CrossRef]
2. Qays, M.O.; Ahmad, I.; Abu-Siada, A.; Hossain, M.L.; Yasmin, F. Key communication technologies, applications, protocols and future guides for IoT-assisted smart grid systems: A review. *Energy Rep.* **2023**, *9*, 2440–2452. [CrossRef]
3. Mehannaoui, R.; Mouss, K.N.; Aksa, K. IoT-based food traceability system: Architecture, technologies, applications, and future trends. *Food Control* **2023**, *145*, 109409. [CrossRef]
4. Saleem, F.; Wyne, S. WLAN-based indoor localization using neural networks. *J. Electr. Eng.* **2016**, *67*, 299–306. [CrossRef]
5. Rahman, M.A.A.; Karim, M.K.; Bundak, C.E. Weighted local access point based on fine matching k-nearest neighbor algorithm for indoor positioning system. In Proceedings of the 2019 AEIT International Annual Conference, Florence, Italy, 18–20 September 2019; pp. 1–5.
6. Gupta, D.; Wadhwa, S.; Rani, S.; Khan, Z.; Boulila, W. EEDC: An Energy Efficient Data Communication Scheme Based on New Routing Approach in Wireless Sensor Networks for Future IoT Applications. *Sensors* **2023**, *23*, 8839. [CrossRef] [PubMed]
7. Khan, A.; Abdeljawad, T.; Alqudah, M.A. Neural networking study of worms in a wireless sensor model in the sense of fractal fractional. *AIMS Math.* **2023**, *8*, 26406–26424. [CrossRef]
8. Ebied, M.; Elmisery, F.A.; El-Hag, N.A.; Sedik, A.; El-Shafai, W.; El-Banby, G.M.; Soltan, E.; Al-Zubi, N.; Brisha, A.; Zekry, A.; et al. A Proposed Deep-Learning-Based Framework for Medical Image Communication, Storage and Diagnosis. *Wirel. Pers. Commun.* **2023**, *131*, 2331–2369. [CrossRef]
9. Realini, E.; Caldera, S.; Pertusini, L. Precise GNSS positioning using smart devices. *Sensors* **2017**, *17*, 2434. [CrossRef] [PubMed]
10. Yeh, S.; Hsu, W.; Lin, W.; Wu, Y. Study on an indoor positioning system using Earth's magnetic field. *IEEE Trans. Instrum. Meas.* **2020**, *69*, 865–872. [CrossRef]
11. Tao, Y.; Wu, L.; Sidén, J.; Wang, G. Monte Carlo-based indoor RFID positioning with dual-antenna joint rectification. *Electronics* **2021**, *10*, 1548. [CrossRef]
12. Scherhäufl, M.; Rudic, B.; Stelzer, A.; Pichler-Scheder, M. A blind calibration method for phase-of-arrival-based localization of passive UHF RFID transponders. *IEEE Trans. Instrum. Meas.* **2019**, *68*, 261–268. [CrossRef]
13. Zhou, J.; Ke, Y.; Yu, K.; Tian, D. A solution of high-precision WLAN positioning based on TDOA and PTP. *MATEC Web Conf.* **2016**, *61*, 7018. [CrossRef]
14. Alshami, I.H.; Ahmad, N.A.; Sahibuddin, S.; Firdaus, F. Adaptive indoor positioning model based on WLAN-fingerprinting for dynamic and multi-floor environments. *Sensors* **2017**, *17*, 1789. [CrossRef] [PubMed]
15. Fang, K.T.; Lee, C.T.; Sun, L. An improved hierarchical WLAN positioning method based on apriori knowledge. *Electronics* **2019**, *8*, 475. [CrossRef]

16. Bencak, P.; Hercog, D.; Lerher, T. Indoor positioning system based on Bluetooth low energy technology and a nature-inspired optimization algorithm. *Electronics* **2022**, *11*, 308. [CrossRef]
17. Chen, J.; Wang, S.; Ouyang, M.; Xuan, Y.; Li, K.C. Iterative positioning algorithm for indoor node based on distance correction in WSNs. *Sensors* **2019**, *19*, 4871. [CrossRef] [PubMed]
18. Benatia, M.A.; Sahnoun, M.; Baudry, D. Multi-objective WSN deployment using genetic algorithms under cost, coverage, and connectivity constraints. *Wirel. Pers. Commun.* **2017**, *94*, 2739–2768. [CrossRef]
19. Li, T.; Yan, W.; Ping, L.; Fang, P. A WSN positioning algorithm based on 3D discrete chaotic mapping. *EURASIP J. Wirel. Commun. Netw.* **2019**, *126*, 1323. [CrossRef]
20. Yanbin, H.; Xiaodong, Y.; Abbasi, Q.H. Efficient AoA-based wireless indoor localization for hospital outpatients using mobile devices. *Sensors* **2018**, *18*, 3698. [CrossRef]
21. De Gante, A.; Siller, M. A survey of hybrid schemes for location estimation in wireless sensor networks. *Procedia Technol.* **2013**, *7*, 377–383. [CrossRef]
22. Abed, A.; Abdel-Qader, I. RSS-fingerprint dimensionality reduction for multiple service set identifier-based indoor positioning systems. *Appl. Sci.* **2019**, *9*, 3137. [CrossRef]
23. Arai, H. Antennas in Access Points of WLAN/WiFi. In *Handbook of Antenna Technologies*; Chen, Z., Liu, D., Nakano, H., Qing, X., Zwick, T., Eds.; Springer: Singapore, 2016; pp. 2579–2588.
24. Ali, R.A.E.Y.; Babiker, A.; Mustafa, A.N. Performance analysis of WiFi network. *IOSR J. Electron. Commun. Eng.* **2015**, *10*, 67–76.
25. Farid, Z. Higher Accuracy Hybrid Wireless Indoor Localization Using Machine Learning. Ph.D. Thesis, Faculty of Engineering, Universiti Kebangsaan Malaysia, Bangi, Malaysia, 2016.
26. Jiang, L. A WLAN Fingerprinting Based Indoor Localization Technique. Master's Thesis, Department Computer Science, University of Nebraska, Lincoln, NE, USA, 2012.
27. Khan, I.U.; Ali, T.; Farid, Z.; Scavino, E.; Rahman, M.A.A. An improved hybrid indoor positioning system based on surface tessellation artificial neural network. *Meas. Control* **2020**, *53*, 1968–1977. [CrossRef]
28. Chen, J.; Song, S.; Yu, H. An indoor multi-source fusion positioning approach based on PDR/MM/WiFi. *AEU—Int. J. Electron. Commun.* **2021**, *135*, 153733. [CrossRef]
29. Luo, J.; Zhang, Z.; Liu, C.; Luo, H. Reliable and cooperative target tracking based on WSN and WiFi in indoor wireless networks. *IEEE Access* **2018**, *6*, 24846–24855. [CrossRef]
30. Guo, T.; Chai, M.; Xiao, J.; Li, C. A hybrid indoor positioning algorithm for cellular and Wi-Fi networks. *Arab. J. Sci. Eng.* **2022**, *47*, 2909–2923. [CrossRef]
31. Gang, H.-S.; Pyun, J.-Y. A smartphone indoor positioning system using hybrid localization technology. *Energies* **2019**, *12*, 3702. [CrossRef]
32. Scavino, E.; Rahman, M.A.A.; Farid, Z. An improved hybrid indoor positioning algorithm via QPSO and MLP signal weighting. *Comput. Mater. Contin.* **2023**, *74*, 379–397. [CrossRef]
33. Farid, Z.; Khan, I.U.; Scavino, E.; Rahman, M.A.A. A WLAN fingerprinting based indoor localization technique via artificial neural network. *IJCSNS Int. J. Comput. Sci. Netw. Secur.* **2019**, *19*, 157–165.
34. Farid, Z.; Nordin, R.; Ismail, M.; Abdullah, N.F. Hybrid indoor-based WLAN-WSN localization scheme for improving accuracy based on artificial neural network. *Mob. Inf. Syst.* **2016**, *2016*, 6923931. [CrossRef]
35. Xiao, P.Y.; Peng, J.; Zhou, G. An improved QPSO algorithm based on social learning and Lévy flights. *Syst. Sci. Control. Eng.* **2018**, *6*, 364–373.
36. Sun, J.; Feng, B.; Xu, W. Particle swarm optimization with particles having quantum behaviour. In Proceedings of the 2004 Congress on Evolutionary Computation, Portland, OR, USA, 19–23 June 2004; pp. 325–331.
37. Bhatia, A.S.; Saggi, M.K.; Zheng, S. QPSO-CD: Quantum-behaved particle swarm optimization algorithm with Cauchy distribution. *Quantum Inf. Process.* **2020**, *19*, 345. [CrossRef]
38. Russell, S.; Norvig, P. *Artificial Intelligence—A Modern Approach*, 4th ed.; Pearson Prentice Hall: Hoboken, NJ, USA, 2020.
39. Kingma, D.P.; Ba, J.L. Adam: A method for stochastic optimization. In Proceedings of the International Conference on Learning Representations (ICLR), San Diego, CA, USA, 7–9 May 2015; pp. 1–13.
40. Ficco, M.; Palmieri, F.; Castiglione, A. Hybrid indoor and outdoor location services for new generation mobile terminals. *Pers. Ubiquitous Comput.* **2014**, *18*, 271–285. [CrossRef]
41. Xiong, Z.; Song, Z.; Scalera, A.; Ferrera, E.; Sottile, F.; Brizzi, P.; Tomasi, R.; Spirito, M.A. Hybrid WSN and RFID indoor positioning and tracking system. *EURASIP J. Embed. Syst.* **2013**, *1*, 6. [CrossRef]

algorithms

MDPI

Article

A PSO-Based Approach for the Optimal Allocation of Electric Vehicle Parking Lots to the Electricity Distribution Network

Marzieh Sadat Arabi † and Anjali Awasthi *,†

Concordia Institute for Information Systems Engineering (CIISE), Concordia University, Montreal, QC H3G 2W1, Canada; marzieh.aarabi1986@gmail.com

* Correspondence: anjali.awasthi@concordia.ca; Tel.: +1-514-848-2424 (ext. 5622)

† These authors contributed equally to this work.

Abstract: Electric vehicles can serve as controllable loads, storing energy during off-peak periods and acting as generation units during peak periods or periods with high electricity prices. They function as distributed generation resources within distribution systems, requiring controlled charging and discharging of batteries. In this paper, we address the problem of the optimal allocation of parking lots within a distribution system to efficiently supply electric vehicle loads. The goal is to determine the best capacity and size of parking lots to meet peak hour demands while considering constraints on the permanent operation of the distribution system. Using the particle swarm optimization (PSO) algorithm, the study maximizes total benefits, taking into account network parameters, vehicle data, and market prices. Results show that installing parking lots could be economically profitable for distribution companies and could improve voltage profiles.

Keywords: electric vehicles; vehicle-to-grid charging; parking lot allocation; particle swarm optimization

1. Introduction

In recent times, various challenges, such as environmental concerns, dwindling fuel resources, volatile fuel prices, and the necessity to reduce reliance on fossil fuels, have led to the recognition of electric vehicles (EVs) as a valuable asset in both transportation and power systems [1]. The concept of a smart grid, as highlighted by [2], revolves around environmental conservation and incorporates elements like renewable energy sources (such as wind and solar power), demand response systems, and distributed generation technologies like EVs. This approach aims to optimize asset utilization, ensure reliable system operation, and offer greater choice to consumers.

The research by Markel and Bennion [3] demonstrates that vehicles are parked for approximately 93–96% of the day, making them available for alternative uses such as serving as storage devices for the grid. This availability, combined with the increasing need for cost-effective energy storage solutions in the power system, suggests that EVs could be utilized as limited energy resources in the power system [4]. In addition to serving as storage devices, EVs can also be used as controllable loads. This means they can be operated as batteries to store energy during off-peak periods and as generation units during peak periods or intervals with high electricity prices. Despite their limited power output, EVs can still be a valuable distributed generation (DG) resource for V2G (vehicle-to-grid) technology implementation in smart grids.

DG technologies offer many economic and technical benefits, as highlighted in the work in [5,6]. However, these benefits can only be maximized when the optimal sizes and locations of DG units are determined. Sound decision-making in this regard can provide benefits to the distribution network, suppliers, and customers alike.

Optimal allocation of parking lots [7–10], as a new type of distributed generation (DG), should be prioritized alongside other resources for electricity generation. One of the key distinctions between parking lots and other traditional distributed generators (DGs) is the stochastic nature of their output. Due to the stochastic nature of EV owners' behavior, the output of parking lots is also stochastic. Parking lots serve as charging stations for EVs for driving purposes. Parking lots function as storage devices, storing electrical power in the batteries of vehicles during times of low electricity prices and delivering power to the distribution system during times of high electricity prices.

High penetrations of distribution-connected storage devices or plug-in vehicles can have adverse impacts on the grid due to their charging loads, which are often randomly located or unmanaged additions. However, the optimal allocation of parking lots can help mitigate these issues by reducing network losses, improving voltage profiles, and consequently bringing economic benefits for the distribution system company (DISCO).

1.1. Research Objectives

The objectives of this research study are as follows:

- To determine the optimal location and size of parking lots to meet the demand of EV owners at peak time;
- To improve the distribution system voltage profile;
- To provide increased economic benefit to the distribution system.

1.2. Related Works

In the literature, various methods have been employed to solve the problem of locating parking lots. These methods can be broadly categorized into optimization, metaheuristics, machine learning, and game theory.

Optimization models are the most popular method used for EV parking lot allocation problems. Algafri et al. [11] addressed smart city charging station allocation for electric vehicles using analytic hierarchy process and multi-objective goal programming. The primary objective is to assign electric vehicles to the most suitable charging stations, taking into account four objectives: maximizing the energy requested by electric vehicles, minimizing total response time, reducing charging costs, and minimizing battery degradation. Several factors, such as travel distance, state of charge, grid-to-vehicle energy trading, time, road traffic, driver priorities, and charging station capacity, are taken into account to address this problem. Tan et al. [12] study a multi-agent system for electric vehicle charging scheduling in parking lots. Casella, Ferro, and Robba [13] propose a decentralized optimization approach to the power management of electric vehicle parking lots.

Metaheuristics are the second commonly reported method to address the parking lot allocation and sizing problem. A model for multi-objective site selection of electric vehicle charging stations based on the Non-Dominated Sorting Genetic Algorithm (NSGA)-II is proposed by Zhang and Shi [14]. A multi-objective neural network algorithm (MONNA) for the optimal location of electric vehicle charging infrastructure in Tunis city is presented in Mehouachi et al. [15]. In El-Zonkoly and Santos Coelho [16], a multi-objective algorithm based on the artificial bee colony (ABC) and firefly algorithms is proposed to optimally determine the number of parking lots to be allocated for plug-in hybrid electric vehicles (PHEVs) in a distribution system is proposed. In Fathy and Abdelaziz [17], Gray Wolf Optimizer for optimal sizing and siting of energy storage systems in electric distribution networks is investigated. Kempton and Tomic [18] present a method based on calculating the probability of electric vehicles (EVs) entering each parking lot for the long-term planning of EV parking lots.

Ahmadi et al. [19] perform optimal allocation of EV parking lots and DG in a micro grid using a two-stage genetic algorithm and particle swarm optimization (GA-PSO). Ferraz et al. [20] developed a multi-objective optimization approach for the allocation of electric vehicle parking lots and smart charging with distributed energy resources. Kumar et al. [21] performs hybrid genetic algorithm-simulated annealing-based electric vehicle charging station placement for optimizing distribution network resilience. Duan et al. [22] proposes an improved metaheuristic method based on fire hawks for the optimization of electric vehicle parking lots in a distribution network. Amini and Islam [23] studies the reliability-constraint allocation of EV parking lots is a radial distribution network. A genetic algorithm is utilized to achieve the best allocation of parking lots. The numerical results show that distribution feeders with no residential and commercial customers are not feasible to allocate parking lots. Shaheen et al. [24] performs optimal electric vehicle charging and discharging scheduling using metaheuristic algorithms and proposes a V2G approach for cost reduction and grid support.

Machine learning models for parking lot allocation have also gained traction in recent years. Ding, Gan, and Shaker [25] perform optimal management of parking lots using big data for electric vehicles via internet of things and Long Short-term Memory. Shariatzadeh et al. [26] address charging scheduling in a workplace parking lot and develop bi-objective optimization approaches through predictive analytics of electric vehicle users' charging behavior. Shahriar et al. [27] provide a review on machine learning approaches for EV charging behavior. Varone et al. [28] models solar parking lot management and proposes an IoT platform for smart charging EV fleets, using real-time data and production forecasts. Pourvaziri et al. [29] develops an integrated deep learning and queueing theory approach for the planning of electric vehicle charging stations. Guo et al. [30] designs a reinforcement learning-based intelligent car transfer planning system for parking lots.

Game theory is the fourth most popular method for EV parking lot allocation and sizing. Adil et al. [31] perform optimal location and pricing of electric vehicle charging stations using machine learning and Stackelberg game. Lv et al. [32] proposes a decision framework for improving the service quality of charging stations based on online reviews and evolutionary game theory. Boateng et al. [33] develops a dynamic pricing and reservation-based framework leveraging multi-agent reinforcement learning in automated valet parking and charging. Abbasi et al. [34] proposes a coupled game theory and Lyapunov optimization approach to electric vehicle charging at fast charging stations. Morais [35] proposes a new approach for electric vehicle charging management in parking lots considering fairness rules. Kempton [36] presents a new approach based on "EV owners' welfare" that is proposed to determine the optimal number, location and capacity of each electric vehicle (EV) parking lot to maximize the profit of electrical distribution companies. Ref. Jannati-Oskuee et al. [37] performed risk-based joint flexible distribution network expansion planning and allocation of EV parking lots using information gap decision theory (IGDT).

In this paper, we address V2G systems and propose a PSO-based approach to address the EV parking lot allocation and sizing problem. PSO is valued for its simplicity, ease of implementation, and its effectiveness in handling non-convex and multi-modal optimization challenges. However, it can sometimes face issues with premature convergence, where particles settle on a suboptimal solution without thoroughly exploring the full search space.

The rest of this paper is organized as follows. In Section 2, the mathematical model is presented. Section 3 presents the solution approach. In Section 4, a case study using the proposed approach is provided. In Section 5, a comparison of model results with other approaches available in the literature is performed. Conclusions and recommendations for future works are presented in Section 6.

2. Mathematical Model

The primary aim of the proposed model is to identify suitable locations and optimal sizes for EV parking lots by maximizing the objective function [34,38]. It consists of 5 cost functions that are homogenized based on weighted values as shown in Equation (1).

2.1. Objective Function

$$MaxF = \sum_{i=1}^{N_{V2G}} [w_1 r(i) - w_2 CF_{cap(i)} - w_3 CF_{(Pu,driving)(i)} - w_4 CF_{(pu,V2G)(i)}] + \sum_{j=1}^{J} w_5 DC_{loss(j)} \quad (1)$$

where i is the set of parking lots, j is the number of load levels, N_{V2G} is the number of parking lots, and $w_1, \ldots, w_5$ are weighting coefficients. Each of the five cost functions are explained in detail as follows.

2.1.1. Charging Time of an EV

The charging time of an electric car depends on several factors, including the battery capacity, the charging power, and the initial state of charge of the battery. The charging time for full charging of an EV as a function of initial state of charge (SOC) can be calculated as Equation (2).

$$t(i) = \frac{(1 - SOC_i) * ES_i}{P_v} \quad (2)$$

where SOC_i is the initial SOC of the vehicle i, ES_i is the battery capacity of EV i, and P_v is the power rate with which EV is charged. We are considering three levels of SOC for batteries, as depicted in Table 1, and the parking lot's input–output power will exhibit three stages, similar to the illustration in Figure 1 [1].

Table 1. Initial SOC of available vehicles in the parking lot.

Initial SOC	Number of Vehicles
SOC1 (0.3)	n1
SOC2 (0.45)	n2
SOC3 (0.7)	n3

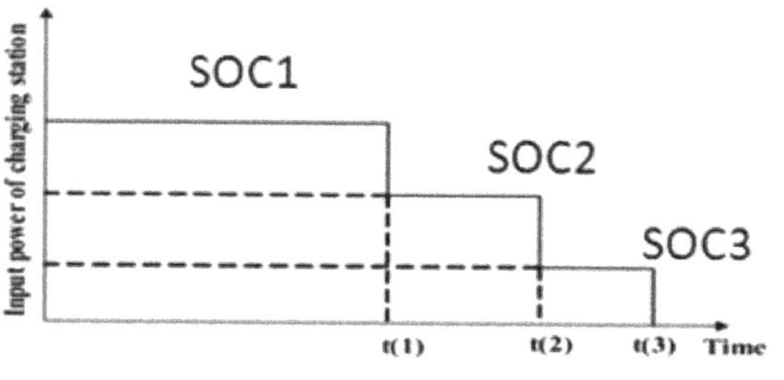

Figure 1. Input–output power modeling of the parking lots.

It is worth noting that the charging process does not always happen at a constant rate. Typically, charging follows a curve where the speed is faster when the battery's SOC is low, and it slows down as the battery approaches full capacity. This adjustment in charging speed is managed by the battery management system to ensure the battery's health and safety are optimized. Moreover, certain electric vehicles are compatible with different charging speeds, including slower AC charging and faster DC fast charging. The charging duration can vary significantly based on the type of charging station and the capabilities of the EV. With advancements in EV technology, enhancements in charging infrastructure, and progress in battery chemistry, it is probable that charging times for electric vehicles will further improve.

2.1.2. Output Power of the EV Parking Lot

The output power of the EV parking lot is equal to

$$P_{parkch} = P_v * n \tag{3}$$

where P_v represents the power rate at which the EV is charged, n represents the number of available vehicles in the parking lot, and P_{parkch} denotes the total electrical power that the parking lot can supply to the connected vehicles. This capacity is determined by the parking lot's charging capacity and the number of charging connectors it possesses. The charging capacity indicates the maximum power that the charging station can deliver at any given moment. This capacity may vary depending on factors such as the type of charging station (AC or DC fast charging), the capabilities of the charging infrastructure, and constraints within the local electrical grid.

For example, if you have a DC fast charging station with a charging capacity of 150 kW and it has two charging connectors, the total output power of the station would be

Output Power = 150 kW × 2 = 300 kW

This means that the station can deliver a maximum of 300 kilowatts of power to the connected electric vehicles simultaneously.

It is important to consider that although the charging station has a specific output power capacity, individual vehicles may not always draw the maximum available power. The charging rate of each EV is influenced by factors such as the capabilities of the vehicle's onboard charger, the battery's state of charge, and the power supply of the charging station. As a result, the actual charging rate of each vehicle may be lower than the station's maximum output power.

2.1.3. Revenues of EVs

The profit made by parking lots from selling charging services to electric vehicles is calculated as

$$r(i) = Pr_p * P_{parkch(i)} * t_{disp(i)} \tag{4}$$

where $r(i)$ is the total revenue gained from the ith parking lot, $t_{disp(i)}$ is the total time that the V2G power is dispatched, and Pr_p is the market price of electricity at peak time.

2.1.4. Required Costs

The cost of vehicle-to-grid (V2G) power consists of three components: purchased energy, wear, and capital costs. The purchased energy and wear costs for V2G are additional expenses required specifically for V2G, not for driving. Similarly, the capital cost represents the expense of additional equipment necessary for V2G, distinct from the primary function of vehicles, which is transportation. In addition to the cost of V2G power, this section also models the cost of purchased energy for driving purposes [39].

1. Capital cost of parking lot:

$$CF_{cap(i)} = C_{ac} * PC(i) \tag{5}$$

C_{ac} is the annualized capital cost for each vehicle. $PC(i)$ is the capacity of parking lot (i).

2. Cost of purchased energy to charge EVs for driving

$$CF_{pu.driving(i)} = \sum_{d=1}^{t_n} \frac{Pr_{off}}{\mu_{conv}} * P_{parkch}(i,d) * t(d) \quad (6)$$

where Pr_{off} is the market price of electricity at off-peak times, μ_{conv} is the efficiency of the inverter, $P_{parkch}(i,d)$ is the required power to charge vehicles from SOC 0 to SOC1 during the duration of $t(d)$ (which is calculated with Equation (3)), and t(d) it is calculated with Equation (2).

3. Cost of purchased EV charging energy to charge vehicles for V2G power

$$CF_{(pu.V2G)}(i) = Pr_{pe} * P_{park(i)} * t_{disp} \quad (7)$$

The value of Pr_{pe} represents the energy cost purchased and is calculated using the following equation:

$$Pr_{pe} = \frac{Pr_{off}}{\mu_{conv}} + C_d \quad (8)$$

where C_d is the cost of equipment degradation due to the extra use for V2G, and μ_{conv} is the efficiency of the inverter.

2.1.5. Loss Mitigation Benefit

The power loss in the distribution system varies due to the output power of parking lots. Therefore, the cost of system loss can be evaluated using the following expression:

$$DC_{loss(j)} = price(j) * (loss(j) - loss_{V2G}(j)) \quad (9)$$

where

$$loss(j) = \sum_{b=1}^{B} R_b * I_b^2(j) * t(k) \quad (10)$$

where $price(j)$ represents the cost of electricity at load level j, $loss(j)$ is the network loss at load level j without V2G, $loss_{V2G}$ is the network loss at load level j with V2G, R_b is the resistance of branch b, and $I_b(j)$ is the current of branch b at time interval j. The time $t(k)$ denotes the required time for full charging of an EV at iteration k.

2.2. Constraints

The objective function is a maximized subject to three inequality constraints, which are described as follows:

1. Distribution line capacity limit. The power flow through the lines must be below the maximum permitted power of the lines due to the line's thermal capacity.

$$S_{(i,j)} \leq S_{(i,j)max} \quad (11)$$

where $S_{(i,j)}$ is the MVA (Mega Volt Amperes) in the line connecting bus i to bus j, and $S_{(i,j)}max$ is the MVA capacity of the line between bus i and bus j.

2. Voltage drops limit. The voltage of each bus should be in the range of minimum and maximum voltages.

$$|V|_{min} \leq |V_{n_b}| \leq |V|_{max} \quad (12)$$

where $|V|_{min}$ and $|V|_{max}$ are the minimum and maximum allowable voltages at buses, respectively.

3. Vehicle limit in each parking lot. The capacity of each parking lot in a specific area is constrained by the number of EVs in that area. This constraint can be expressed as follows:

$$CP \leq CP_{max} \tag{13}$$

where CP_{max} is the maximum capacity of a parking lot, which can be installed.

3. Solution Approach

Figure 2 presents the various steps of the proposed solution approach. In the first step, data are collected for the network, EVs, technical parameters, and cost. The AHP (analytic hierarchy process) method is then used to determine the weight of the costs and the revenue criteria. Following this, the backward/forward sweep power flow method is used to calculate power losses in the distribution network. The particle swarm optimization (PSO) algorithm is then employed to determine the number of electric vehicles in each bus and reduce the losses. Finally, the revenue is calculated, which is net income minus the required cost, and at the end, the total benefit is computed. The model was built using Python.

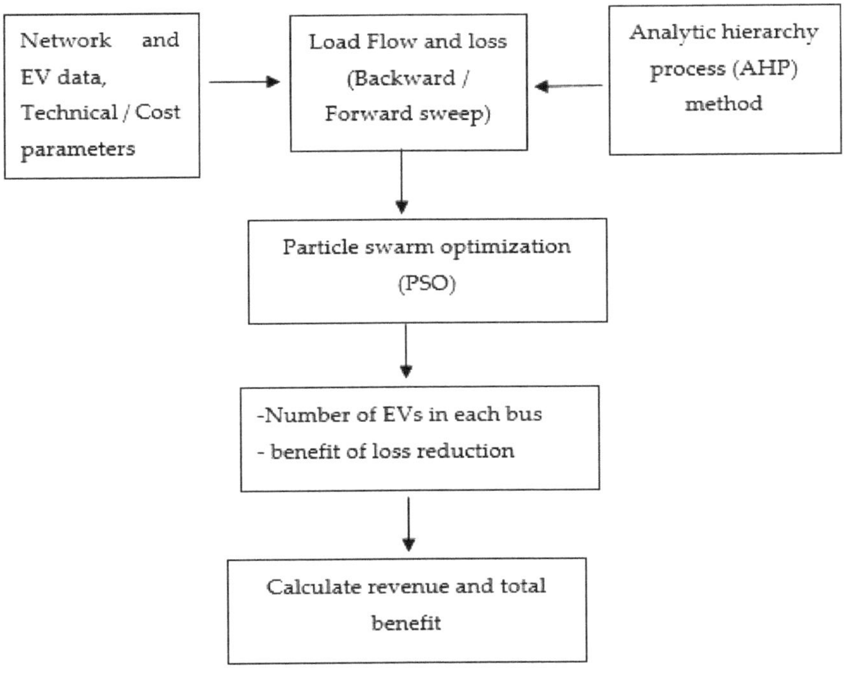

Figure 2. Different phases in the suggested solution approach.

3.1. Assumptions

In order to allocate parking lots, some assumptions [40–42] are taken into account as follows:

- This study assumes that the distribution company is responsible for supplying customer demand, installing parking lots, and controlling the charging and discharging of EV batteries. DISCO aims to fulfill these responsibilities while minimizing costs and improving the quality and reliability of customer service.

- It is important to note that in calculating profits, the study assumes that DISCO does not receive compensation from EV owners for battery charging required for driving purposes. Additionally, any vehicle degradation costs due to vehicle-to-grid (V2G) operations are covered by DISCO and paid to EV owners. These assumptions are made to incentivize EV owners to park their vehicles in the parking lot during days with high-priced peak electricity.
- All vehicles are assumed to be charged and discharged at their maximum charging rate. It is important to note that this assumption is common in several EV studies.
- In the modeling of parking lots, it is assumed that the initial state of charge of EVs has three levels (Figure 1). However, the proposed model can be adapted for use with other SOC levels as well. The initial SOC of vehicles can be adjusted using a suitable distribution function, and parking lots can be optimally placed considering this function.
- Another assumption made in the modeling of parking lots is that all batteries are the same size. As a result, the output power of the parking lot is assumed to be constant during the discharging state.
- Under traditional approximations used by utilities, there might be 200 peak hours in a year during which an incremental kW h of electricity would be worth USD 0.5/kW h [18]. Therefore, the maximum hours that vehicles deliver power to the network is assumed to be 200 h in a year.

3.2. *Load Flow*

Load flow studies [43] are essential for ensuring stable, reliable, and economical transmission of electrical power from generators to consumers through the grid system. We refer to the movement of both reactive and active power as "flow" or "load flow". A mathematical technique for determining distinct bus voltages, their phase angles, and the flow of reactive and active power across various branches, generators, and loads in a steady-state setting is provided by power flow studies. The backward/forward sweep method is used to evaluate load flow in radial distribution. First, voltage data on the main line and its derivatives are all adjusted to the nominal voltage. Next comes an iterative procedure where Kirchhoff's Current Law (KCL) is used to compute the branch current in the main line in a backward sweep after currents in derivative lines have been determined. Then, node voltages are calculated using Kirchhoff's Voltage Law (KVL) in a forward sweep. Until the voltage magnitudes at each node in the current iteration and the previous iterations are lower than the tolerance limit, this backward and forward sweep approach is repeated.

3.2.1. Backward Sweep

This update moves towards the branches that are related to the source node from the branches that are the farthest away from it. During the backward propagation phase, the updated load flows in each branch are computed while taking the node voltages from the previous iteration into account. The voltage levels determined in the forward propagation phase are maintained throughout this procedure. Next, using the backward propagation method, the modified power loads in each branch are propagated backward along the feeder. This indicates that the backward propagation proceeds in the direction of the source node, beginning at the node that is farthest away from it.

3.2.2. Forward Sweep

The forward sweep in the backward/forward sweep technique entails updating nodal voltages from first-layer branches to last-layer branches. Computing the voltages at each node, starting with the feeder source node, is the aim of forward propagation. At the feeder substation, the voltage is adjusted to reflect its true value. The effective power in each branch is kept constant during the forward propagation phase at the value determined during the backward propagation step. The main steps of the suggested method, together with the relevant equations, are listed below.

Step 1: Initialization of voltages V^0_{node}(assume an initial voltage at all nodes of $n = 2, 3 \ldots N$, where N is the number of nodes).

$$V^0_{node} = V_{line} < 0 \tag{14}$$

Step 2: Iteration count initialization, K = 1. convergence_threshold = tolerance $\epsilon = 10^{-5}$.

Step 3: Load Current I^k_{node} computation.

$$I^k_{node} = \frac{S^{(k-1)}_{node}}{(v^{(k-1)}_{node}} \tag{15}$$

For node = 2, 3, …., N. s represents the complex power at the load and v is the voltage. $s = p + jQ$ (in the simulation, the parking lot is modeled as a bus, $Q = 0$).

Step 4: Backward sweep: update the current starting from the end nodes.

$$I_{branch} = I^k_{node} + \sum I_{branch,downstream} \tag{16}$$

Step 5: Forward sweep (start from the substation and move towards the end nodes).

$$V^k_{node} = V^k_{previousnode} - (Z_{branch} * I^k_{branch}) \tag{17}$$

For all node = 2, 3, …, N. $Z_{branch} = R_{branch} + jX_{branch} = \sqrt{(R^2 + X^2)}$, where R is the resistance of each branch and X is the reactance of each branch.

Step 6: Error.

$$E^k_{node} = |V^k_{node} - V^{(k-1)}_{node}|/V_{line} \leq tolerance \tag{18}$$

for node = 2, 3…., N.

Step 7: Maximum error.

$$e_{max} = max(E^k_2, E^k_3, E^k_4, \ldots\ldots, E^k_{node}) \tag{19}$$

Step 8: The load flow is converging if e_{max} is less than or equal to tolerance (ϵ). If not, proceed to Step 3 and repeat the process after updating the iteration count to K = K + 1. The operation of the load flow computation utilizing the backward and forward sweep approach is shown in Figure 3.

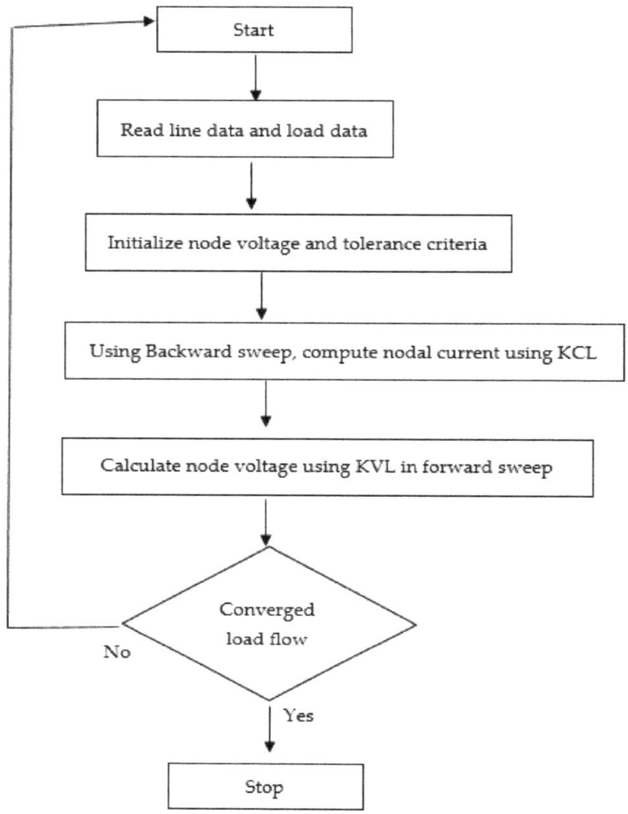

Figure 3. Load flow calculation using backward/forward sweep method.

3.3. *Particle Swarm Optimization (PSO)*

PSO [44] is a nature-inspired optimization technique that draws inspiration from the collective behavior of groups in nature, like the flocking of birds or the schooling of fish. PSO works by simulating the movement of a swarm of particles through a multi-dimensional search space, aiming to find the best solution for a given problem. It is particularly effective for solving optimization problems with numerous variables and complex, non-linear objective functions. The objective function stated in Equation (1) is maximized with the PSO (particle swarm optimization) algorithm. Each member of the group in PSO is characterized by a velocity vector and a position vector in the search space. With each iteration, the particles' new positions are determined based on their velocity and current position in the search space.

In the problem's dimensional space, solutions are represented by matrices with corresponding dimensions. Therefore, the position and velocity of PSO can be denoted by matrices of size $X_i = [x_{i,1}, x_{i,2}, , x_{i,D}]$ and $[V_i = v_{i,1}, v_{i,2}, , v_{i,D}]$, respectively, where $x_{i,j}$ represents the value of position x_i in dimension j and $V_{i,j}$ represents the value of velocity V_i in dimension j. The operations of the algorithm are also conducted as matrix operations, and the velocity and position update models of the standard PSO are represented as follows:

$$V_{i,j}^{t+1} = w * (Vmax_{i,j}^{t}) + c1 * r1 * (gBest_j^t - x_{i,j}^t) + c2 * r2 * (pBest_{i,j}^t - x_{i,j}^t) \quad (20)$$

$$x_{i,j}^{t+1} = x_{i,j}^{t} + V_{i,j}^{t+1} \quad (21)$$

$$w = w_{Max} - 1 * \frac{(w_{Max} - w_{Min})}{iters} \quad (22)$$

where $j = 1, 2, ...D$ and $c1$ and $c2$ are constants representing the acceleration coefficients that control the particle's movement. Parameters $r1$ and $r2$ are random numbers following a Gaussian distribution and taking values between 0 and 1, and t represents the current iteration number. $gBest_j^t$ represents the value of the jth dimension of the globally optimal particle in the previous t iterations, and $pBest_{i,j}^t$ represents the value of the jth dimension of the optimal position of particle i in the previous t iterations. The inertia weight parameter, or "w", is a term used in optimization methods like PSO. The inertia weight ('w') in PSO maintains a balance between the local and global search space exploration. Throughout the optimization process, Equation (22) modifies 'w' to progressively lower its value over iterations, which is a popular strategy to balance exploration and exploitation (w_{Max} is the maximum inertia weight, w_{Min} is the minimum inertia weight, and t is the number of iterations).

The algorithm involves four steps:

- Initialization: Initialize a population of particles with random positions and velocities in D dimensions in the search space.
- Estimation: Estimate the fitness of each particle in this population.
- Update: Calculate the speed of each particle and move to the next position.
- Termination: Stop the algorithm if it reaches a certain stop criterion; otherwise, go back to the estimation stage.

The PSO algorithm relies on the interplay between particles' individual experiences (pBest) and the collective best experience of the swarm (gBest) to navigate the search space effectively and converge towards promising regions. The parameters used in the PSO study are V_{max} = 4; w_{Max} = 0.9; w_{Min} = 0.2; c1 = c2 = 2; dim = 2; and PopSize = 5.

3.4. Calculation of Weighting Factors

The analytic hierarchy process (AHP) is used to calculate the weights of the various cost functions of the objective function (Equation (1)). In the first step of AHP, the research factors must be extracted from various sources or obtained from experts. Once the factors and options are identified, the problem is divided into criteria levels and sub-criteria, and arranged in a hierarchical diagram (main criteria at the top). Then, a matrix of pairwise comparisons is formed. The elements of each level are compared pairwise with other related elements at a higher level, and matrices of paired comparisons are created. To assess the importance and preference in these pairwise comparisons, a scale of one to nine is commonly used (Table 2).

Table 2. Preference values for pairwise comparisons.

Numerical Values	Preferences
9	Completely more vital or completely more desired
7	Very strong preference or importance
5	Strong preference or importance
3	A little more important
1	Equal preference or importance
2, 4, 6, 8	Preferences between the above intervals

4. Case Study

Figure 4 shows the case study's test arrangement. Dispersed generations with power factors of 0.9 lag and ranging from 1 to 5 MW have been regarded as negative loads in order to assess the proposed method. Eight load points are powered by a high-voltage distribution substation with a rating of 132–33 kV that is part of the distribution test network. The network's isolator switches, which have a maximum capacity of 25 MVA, isolate each branch. Every load point in the network has a power factor of 0.9 latency, making them all suitable sites for parking lot installations. The test network's technical specifications are shown in Table 3, which loads data in three stages [45].

Table 3. Technical characteristics of branches and load data.

Section from–to	R (Ω)	X (Ω)	L (km)	Load Low Level 1 (MW)	Load Medium Level 2 (MW)	Load High Level 3 (MW)
1–3	1.4	1.5	1.5	5	6	8
3–7	2.78	1.5	1.5	7.5	8.8	9.2
1–2	2	4	4	8.3	11.2	9
2–6	2.8	5.5	5.5	4	5	7
1–5	1.7	1.7	1.7	7.5	8.8	9.2
5–9	2.1	4	4	7.3	10.2	8
1–4	2.26	4.5	4.5	6	7	9
4–8	2.4	5	5	7.5	8.7	9.2

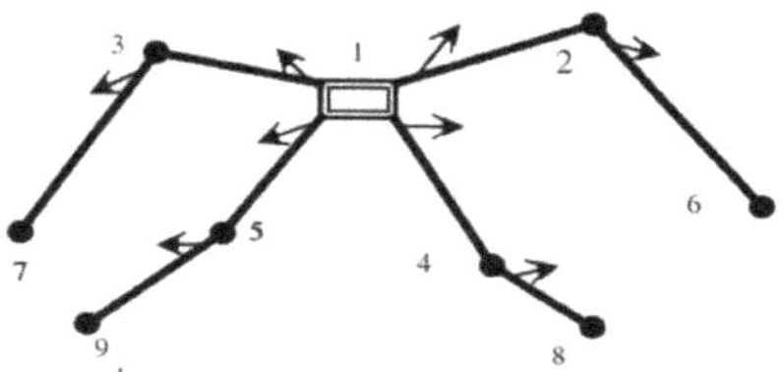

Figure 4. Studied test network.

Table 4 provides a summary of the system's further technical and financial specifications. The following presumptions [18,34] are made in our study:

- The highest billing rate is applied to each vehicle.
- Three degrees of load conditions—light, medium, and peak load—are taken into consideration in this paper.
- The parking lot is represented in the simulation as a bus (Q = 0).
- It is considered that there are always vehicles available.
- Three degrees of initial state of charge (SOC) for electric vehicles (EVs) are assumed in parking lot modeling.
- Every battery has identical dimensions. As a consequence, the output power of the parking lot is continuous in a flat discharging condition.

According to estimates by traditional utilities, there may be 200 or so peak hours annually during which an extra kWh of power is worth 0.5 USD/kWh. As a result, automobiles are believed to provide the network with electricity for no more than 200 h annually.

Table 4. Technical/cost parameters.

Parameters	Values
P_v (Kw)	15, 50
ES (Kwh)	50, 100
t_{disp} (hours/year)	200
$Pr_p(PeakPrice)$ USD kWh	0.5
C_{ac} USD/year for each vehicle	304
Car availability	100%
SOC levels	0.3, 0.45, 0.7
Number of Vehicles per SOC	0.25, 0.25, 0.5
Number of parking lots	3
Pr_{off} USD/kWh	0.05
μ_{conv}	0.85
C_d	0.225
Number of load levels	3
V_{line}	33 kv
V_{min}	0.9 pu
V_{max}	1.1 pu
Price (per load level) USD/kWh	Light load 0.035 Medium load 0.049 Peak load 0.07
t_k (Time duration) h/year	Light load 2190 Medium load 4745 Peak load 1825

4.1. Calculation of Criteria Weights

The AHP approach is used to determine the ideal weighting factors in every scenario. The vector of weighting coefficients is the same in all scenarios and is given by w = [0.2, 0.2, 0.2, 0.2, 0.2].

In the next step, the location of the parking lot will be determined using load flow, and with the PSO algorithm, the number of EVs in each parking lot will be discovered.

4.2. Calculation of Load Flows

The load flow is calculated for four lines (1–3–7, 1–5–9, 1–4–8, and 1–2–6). We will demonstrate the application of the forward/backward sweep technique for line 1–3–7 as an example.

Step 1: Initialization of voltages in eight nodes, $V_2^0 = V_3^0 = V_4^0 = V_5^0 = V_6^0 = V_7^0 = V_8^0 = V_9^0 = V_{line} = 33$ Kv.

Step 2: Iteration count initialization, $k = 1, \epsilon = 10^{-5}$.

Step 3: Load Current computation,

$I_7^1 = \frac{S_7^0}{v_7^1} = \frac{9.2(MW)}{33(KV)} = 278.78$

$I_3^1 = \frac{S_3^0}{v_3^0} = \frac{8(MW)}{33(KV)} = 242.42.$

Step 4: Update current starting from the end nodes (backward sweep),

$I_(3-7)^1 = I_7^1 + \sum_{branch,downstream} = 278.78 + 0 = 278.78$

$I_(1-3)^1 = I_3^1 + \sum_{branch,downstream} = I_3^1 + I_{(3-7)}^1 = 242.42 + 278.78 = 524.21.$

Step 5: Forward sweep,

$V_3^1 = V_{line}^1 - Z_{1-3} * I_{1-3}^1 = 33,000 - 2.051 * 524.21 = 31,930.56$

$V_7^1 = V_3^1 - Z_{3-7} * I_{3-7}^1 = 31,930.56 - 6.16 * 278.78 = 30,212.48.$

Step 6: Error,

$E_3^1 = \frac{|V_3^1 - V_3^0|}{Vline} <= tolerance$

$E_3^1 = \frac{|31,930.56 - 33,000|}{33,000} = 0.0324 > 10^{-5}$

$E_7^1 = \frac{|V_7^1 - 7_7^0|}{Vline} <= tolerance$
$E_7^1 = \frac{|30{,}212.48 - 33{,}000|}{33{,}000} = 0.0844 > 10^{-5}.$
Step 7: Maximum error,
$e_{max} = max(E_3^1, E_7^1) = 0.0844 > 10^{-5}.$
Step 8: If e_{max} is more than tolerance (ϵ), then update the iteration count to k = k + 1 and go to Step 3 and repeat the steps. K = 2.
Iteration 2
Step 3: Load Current computation (line 1-3-7),
$I_7^2 = (S_7^1)/(v_7^1) = \frac{9.2(MW)}{30212.48} = 304.51$
$I_3^2 = (S_3^2)/(v_3^2) = \frac{8(MW)}{31930.56} = 250.54,$
Step 4: Update the current starting from the end nodes (backward sweep),
$I_{3-7}^2 = I_7^2 + \sum_{branch,downstream} = 304.051 + 0 = 304.051$
$I_{1-3}^2 = I_3^2 + \sum_{branch,downstream} = I_3^2 + I_{3-7}^2 = 250.54 + 304.51 = 555.05.$
Step 5: Forward sweep,
$V_3^2 = V_{line}^2 - Z_{1-3} * I_{1-3}^2 = 33{,}000 - 2.051 * 555.05 = 31{,}861.12$
$V_7^2 = V_3^2 - Z_{3-7} * I_{3-7}^2 = 31{,}861.12 - 6.16 * 304.052 = 29{,}984.53.$
Step 6: Error,
$E_3^2 = |V_3^2 - V_3^1|/V_{line} \leq tolerance$
$E_3^2 = |31{,}861.12 - 31{,}930.56|/33{,}000 = 0.021 > 10^{-5}$
$E_7^2 = |V_7^2 - 7_7^1|/V_{line} \leq tolerance$
$E_7^2 = |29{,}984.53 - 30{,}212.48|/33000 = 0.0069 > 10^{-5}.$
Step 7: Maximum error,
$e_{max} = max(E_3^2, E_7^2) = 0.0069 > 10^{-5}.$
Step 8: e_{max} is greater than tolerance (ϵ), so update the iteration count to $k = k + 1$ and go to Step 3. These steps are repeated, and in K = 6, the e_{max} becomes less than tolerance (ϵ), so the load flow is converged at iteration 6.
$e_{max} = max(E_3^6, E_7^6) = 4.28 * 10^{-7} > 10^{-5}$

4.3. Calculating the Location of Parking Lots and Sizing

The location of parking lots, the number of EVs, and the benefit of loss reduction (Equation (10)) are obtained using the PSO algorithm. At the end, the benefit of providing peak power (net income - required cost) and total benefit is calculated using Equation (1).

Additionally, in order to see the voltage enhancement in the presence of V2G, the voltage profile is examined for three different scenarios, both with and without V2G power.

4.4. Scenario Generation

4.4.1. Scenario 1: P_v: 15 kw and ES: 50 Kwh

The outcomes of scenario 1 are presented in Table 5. It is considered that vehicles' availability is 100%. Put otherwise, all EV owners respect the agreement. The total yearly benefit is USD 188,089.803.

Table 5. Simulation results of scenario 1.

Bus number	6	8
Optimum number of EVs	249	302
Benefit of loss reduction (USD)	USD 62,266.4	
Benefit of providing peak power (USD)	USD 125,823.403	
Total benefit (USD)	USD 188,089.803	

Figure 5 displays the voltage profile of load locations during peak hours, when parking lots provide electricity to the distribution system. The algorithm converges at a total of six iterations with a tolerance of 10^{-5} p.u.

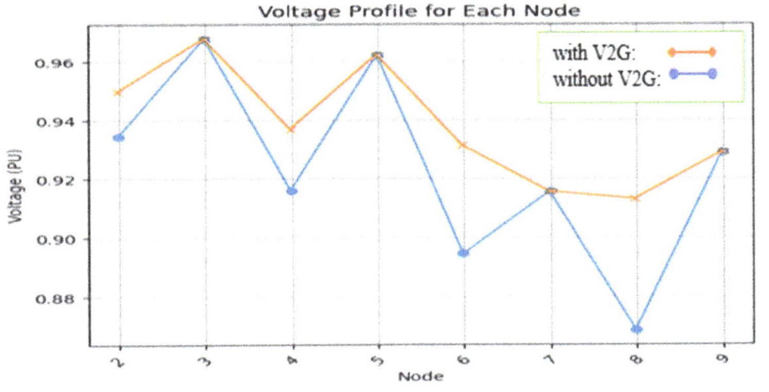

Figure 5. Voltage profile in peak load (scenario 1).

There is a voltage decrease on buses 6 and 8, as seen in Figure 5 and Table 6 (without V2G). Since the voltage on buses 6 and 8 is less than 0.9 p.u., the optimization constraint is not in the required range in this scenario. Nonetheless, when V2G power is present, the voltage profile of the buses improves. It is important to remember that the voltage in every bus is the same as the voltage in every node determined by the load flow technique (after convergence).

Table 6. Voltage magnitude of 9-bus system (50 Kw).

Bus Number	Bus Voltages Without V2G (p.u.)	Bus Voltages with V2G (p.u.)
1	1	1
2	0.9343	0.9496
3	0.9676	0.9676
4	0.9158	0.9368
5	0.9620	0.9620
6	0.8946	0.9311
7	0.9155	0.9155
8	0.8689	0.9130
9	0.9288	0.9288

4.4.2. Scenario 2: P_v: 50 kw and ES: 100 Kwh

The vector of weighting coefficients is calculated like scenario 1. w = [0.2, 0.2, 0.2, 0.2, 0.2]. Table 7 displays the outcome of scenario 2. Vehicle availability is assumed to be 100%. Or, to put it another way, every EV owner honors the agreement. The statistics show that the total yearly benefit is USD 185731.7176.

Table 7. Simulation results of scenario 2.

Bus Number	6	8
Optimum number of EVs	58	74
Benefit of loss reduction (USD)	66,483.2	
Benefit of providing peak power (USD)	119,248.5176	
Total benefit (USD)	185,731.7176	

Figure 6 displays the voltage profile of load points, or parking lots that provide electricity to the distribution system, during peak hours. The total number of iterations needed is six, and the tolerance is 10-5 p.u. Bus numbers 6 and 8 experience a voltage loss when there is no V2G in the network, as shown in Figure 6 and Table 8, since their voltage is less than 0.9 p.u. But when V2G is present, this problem is resolved.

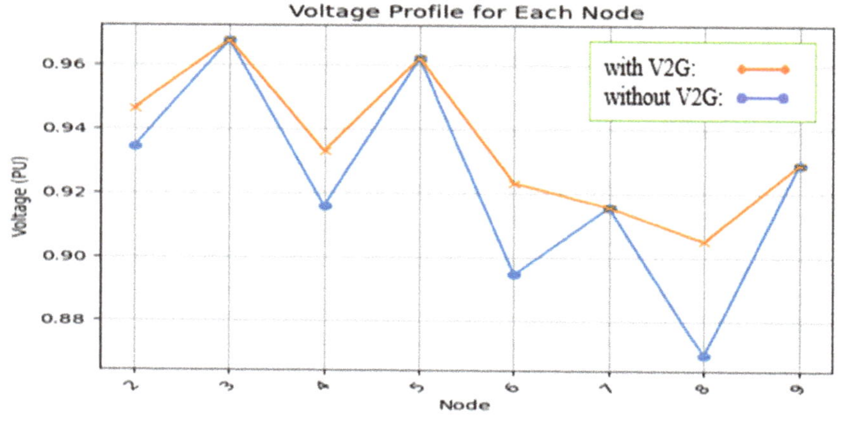

Figure 6. Voltage profile in peak load (scenario 2).

Table 8. Voltage magnitude of 9-bus system (50 Kw).

Bus Number	Bus Voltages Without V2G (p.u.)	Bus voltages with V2G (p.u.)
1	1	1
2	0.9343	0.9462
3	0.9676	0.9676
4	0.9158	0.9329
5	0.9620	0.9620
6	0.8946	0.9229
7	0.9155	0.9155
8	0.8689	0.9049
9	0.9288	0.9288

4.4.3. Scenario 3: P_v: 100 kw and ES: 100 Kwh

w = [0.2, 0.2, 0.2, 0.2, 0.2]

Table 9 displays the results from scenario 3 with the yearly total benefit value as USD 176,902.4823. The availability of vehicles is 100%. Stated differently, every owner of an EV abides by the agreement.

Table 9. Simulation results of scenario 3.

Bus Number	6	8
Optimum number of EVs	26	32
Benefit of loss reduction, USD	68,552.4	
Benefit of providing peak power, USD	108,350.0823	
Total benefit, USD	176,902.4823	

Figure 7 displays the voltage profile of load sites during peak hours when parking lots provide electricity to the distribution system. The total number of iterations needed is six, and the tolerance is 10^{-5} p.u.

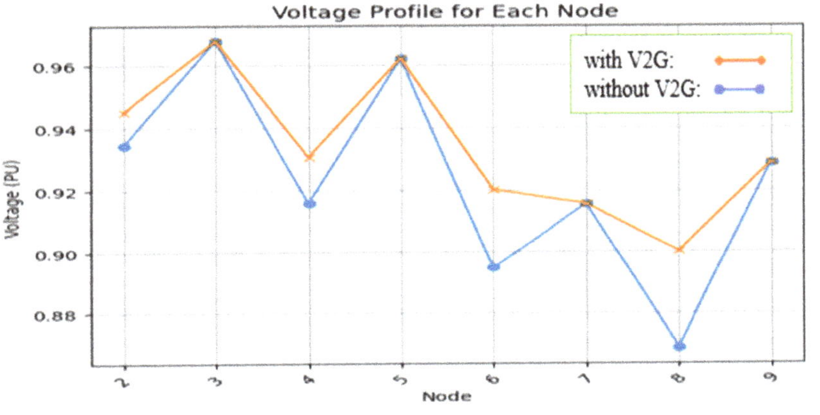

Figure 7. Voltage profile in peak load (scenario 3).

Table 10 and Figure 7 show that there is a voltage reduction at bus numbers 6 and 8, where the voltage is less than 0.9 p.u. (without V2G). In the presence of V2G power, the voltage profile of the buses improves, but the optimization constraints are still within a reasonable range.

Table 10. Voltage magnitude of 9-bus system (100 Kw).

Bus Number	Bus Voltages Without V2G (p.u.)	Bus Voltages with V2G (p.u.)
1	1	1
2	0.9343	0.9449
3	0.9676	0.9676
4	0.9158	0.9306
5	0.9620	0.9620
6	0.8946	0.9200
7	0.9155	0.9155
8	0.8689	0.9001
9	0.9288	0.9288

5. Results Analysis and Discussion

Table 11 presents the comparison results of our work with [44–46]. The outputs are presented in terms of the total numbers of EVs used, voltage drop points (with V2G), total capacity of the EV to the network, and the total benefit of loss reduction. The authors in [34] estimated and located 975 EVs in the network with a total capacity of 14,625 kW. In contrast, ref. [46] explored four methods (Competition over resource optimization—COR, Gray Wolf Optimization—GWO, Water Cycle Algorithm—WCA, and Whale Optimization Algorithm—WOA approaches), resulting in the location of 600, 687, 735, and 712 EVs in the network, respectively, with the same total capacity of 14,625 kW. On the other hand, ref. [47] did not consider EVs in the distribution network; instead, they estimated and located about 5 MW of DGs in the network, distributed across three load types (light, medium, peak). Consequently, their approach demonstrates the potential for significantly reducing losses and increasing benefits compared to others. This study achieves a higher benefit in loss reduction with a smaller number of EVs compared to [34,46]. In addition, the results in [34,46] are based on specific parameters, such as Pv = 15 kW and ES = 50 kWh, without considering other network conditions. Furthermore, the optimization constraint for bus number 8 in [34,46] is not within the appropriate range (voltage drop occurs at bus number 8 during peak hours). This paper, by considering various situations, demonstrates that utilizing fast charging stations and vehicles with higher battery capacities can significantly reduce the total number of EVs while improving voltage drop during peak times.

Table 11. Results comparison.

Methodology	Network Condition	Bus Number	Optimum Number of EVs in Each Bus	Voltage Drop Points (with V2G)	Total Capacity of Network (W)	Benefit of Loss Reduction (USD)
Proposed approach (PSO)	Peak load P_v: 15 kw ES: 50 kwh (200 h)	6 8	249 302	-	8265 KW	62,266.4
	P_v: 50 kw ES: 100 kwh	6 8	58 74	-	6600 KW	66,483.2
	P_v: 100 kw ES: 100 kwh	6 8	26 32	-	5800 KW	68,552.4
Competition over resource optimization algorithm [46]	Peak load (200 h) P_v: 15 kw ES: 50 kwh	2 3 6	100 300 200	8	14,625 KW	33,010
Gray wolf optimizer [46]	Peak load (200 h) P_v: 15 kw ES: 50 kwh	2 3 6	107 350 230	8	14,625 KW	34,110.38
Water cycle algorithm [46]	Peak load (200 h) P_v: 15 kw ES: 50 kwh	2 3 6	115 370 250	8	14,625 KW	38,664.45

Table 11. *Cont.*

Methodology	Network Condition	Bus Number	Optimum Number of EVs in Each Bus	Voltage Drop Points (with V2G)	Total Capacity of Network (W)	Benefit of Loss Reduction (USD)
Whale optimization algorithm [46]	Peak load (200 h) P_v: 15 kw ES: 50 kwh	2 3 6	112 360 240	8	14,625 KW	38,436.149
Genetic algorithm [9]	Peak load (200 h) P_v: 15 kw ES: 50 kwh	2 3 6	375 375 225	8	14,625 KW	38,705
Dynamic programming [45]	Light load (2190 h)	6, 7, 8	-	-	5 MW	1,685,881
	Medium load (4745 h)	6, 7, 8	-	-	5 MW	2,792,897
	Peak load (1825 h)	6, 8	-	-	5 MW	679,784

6. Conclusions and Future Works

This paper aims to address the optimal allocation and sizing of parking lots within a distribution system to efficiently supply loads. A particle swarm optimization-based approach is proposed. The model focuses on maximizing total benefits, considering network data and market prices. Results from the study showed that installing parking lots could be economically profitable for the distribution company and could improve the voltage profile. The simulation results indicate that changes in the battery capacity of EVs in parking lots and the power rate at which the EVs are charged (Pv) result in variations in the outcomes. To achieve more accurate results, it is essential to precisely determine the size of the batteries and the power rate at which the EVs are charged. This would involve careful consideration of factors such as the expected usage patterns of the EVs, the charging infrastructure available, and the overall objectives of the optimization process. Fine-tuning these parameters can lead to more precise and effective outcomes in the optimization of parking lot allocations for EV charging. In addition to the benefits already mentioned, this approach offers several other advantages such as improvements in the voltage profile, reductions in the power flow, and enhanced equipment lifespan. Several avenues for future research and development can be pursued based on the findings and limitations of this work. These directions include the following:

1. A simple system with nine buses was considered in this paper. The proposed method can be used for a complex system like one with 33 buses or 69 buses.
2. The weighting coefficient can lead to variations in the results. In this study, the weighting coefficient was equal and all arrays in matrix A were 1. The same approach can be used to examine the effects of various weighting coefficients.
3. It was assumed that all EV owners respected the contract, and the availability of vehicles was 100 percent. If the availability of vehicles is 80 percent or less, it can have an effect on the result. Reliability improvement has not been investigated in this work, and can be part of future works.

4. In the modeling of parking lots, it is assumed that the initial state of charge (SOC) of EVs has three levels. The proposed model can be used for other SOC levels.

Several operational constraints that can affect the efficiency of the proposed parking system for BEVs/PHEVs, such as ambient temperature, technical issues, or the need for a vehicle to leave the parking lot earlier than planned, can also be integrated in future studies.

Author Contributions: Methodology, M.S.A.; Writing—original draft, M.S.A.; Writing—review & editing, A.A.; Supervision, A.A. All authors have read and agreed to the published version of the manuscript.

Funding: This research received no external funding.

Data Availability Statement: The original contributions presented in this study are included in the article. Further inquiries can be directed to the corresponding author.

Conflicts of Interest: The authors declare no conflicts of interest.

References

1. Moradijoz, M.; Parsa Moghaddam, M.; Haghifam, M.R.; Alishahi, E. A multi-objective optimization problem for allocating parking lots in a distribution network. *Int. J. Electr. Power Energy Syst.* **2013**, *46*, 115–122. [CrossRef]
2. Gellings, C.W. *The Smart Grid: Enabling Energy Efficiency and Demand Response*; River Publishers: Nordjylland, Denmark, 2009.
3. Markel, T.; Bennion, K. *Field Testing Plug-in Hybrid Electric Vehicles with Charge Control Technology in the Xcel Energy Territory*; Technical Report; National Renewable Energy Laboratory (NREL): Golden, CO, USA, 2009.
4. Peterson, B.; Whitacre, J.F. The economics of using plug-in hybrid electric vehicle battery packs for grid storage. *J. Power Sources* **2010**, *195*, 2377–2384.
5. Moradi, M.; Abedini, M. A combination of genetic algorithm and particle swarm optimization for optimal DG location and sizing in distribution systems. *Int. J. Electr. Power Energy Syst.* **2012**, *34*, 66–74.
6. Aman, M.; Jasmon, J.; Mokhlis, H.; Bakar, A. Optimal placement and sizing of a DG based on a new power stability index and line losses. *Int. J. Electr. Power Energy Syst.* **2012**, *43*, 1296–1304.
7. Mirzaei, M.J.; Kazemi, A.; Homaee, O. A probabilistic approach to determine optimal capacity and location of electric vehicles parking lots in distribution networks. *IEEE Trans. Ind. Inform.* **2015**, *12*, 1963–1972.
8. Mohsenzadeh, A.; Pazouki, S.; Ardalan, S.; Haghifam, M.R. Optimal placing and sizing of parking lots including different levels of charging stations in electric distribution networks. *Int. J. Ambient. Energy* **2018**, *39*, 743–750. [CrossRef]
9. Moradijoz, M.; Ghazanfarimeym, A.; Moghaddam, M.P.; Haghifam, M.R. Optimum placement of distributed generation and parking lots for loss reduction in distribution networks. In Proceedings of the 2012 17th Conference on Electrical Power Distribution, Tehran, Iran, 2–3 May 2012; pp. 1–5.
10. Sattarpour, T.; Farsadi, M. Parking lot allocation with maximum economic benefit in a distribution network. *Int. Trans. Electr. Energy Syst.* **2017**, *27*, e2234.
11. Algafri, M.; Alghazi, A.; Almoghathawi, Y.; Saleh, H.; Al-Shareef, K. Smart City Charging Station allocation for electric vehicles using analytic hierarchy process and multiobjective goal-programming. *Appl. Energy* **2024**, *372*, 123775. [CrossRef]
12. Tan, M.; Zhang, Z.; Ren, Y.; Richard, I.; Zhang, Y. Multi-agent system for electric vehicle charging scheduling in parking lots. *Complex Syst. Model. Simulation* **2023**, *3*, 129–142.
13. Casella, V.; Ferro, G.; Robba, M. A decentralized optimization approach to the power management of electric vehicles parking lots. Sustain. *Energy Grids Netw.* **2024**, *38*, 101301.
14. Zhang, H.; Shi, F. A multi-objective site selection of electric vehicle charging station based on NSGA-II. *Int. J. Ind. Eng. Comput.* **2024**, *15*, 293–306.
15. Mehouachi, I.; Trojette, M.; Grayaa, K. MONNA: Multi-objective neural network algorithm for the optimal location of electric vehicle charging infrastructure in Tunis city. *J. Clean. Prod.* **2023**, *431*, 139837. [CrossRef]
16. El-Zonkoly, A.; dos Santos Coelho, L. Optimal allocation, sizing of PHEV parking lots in distribution system. *Int. J. Electr. Power Energy Syst.* **2015**, *67*, 472–477.
17. Fathy, A.; Abdelaziz, Y.A. Grey Wolf Optimizer for Optimal Sizing and Siting of Energy Storage System in Electric Distribution Network. *Electr. Power Components Syst.* **2017**, *45*, 601–614.
18. Kempton, W.; Tomic, J. Vehicle-to-grid power fundamentals: Calculating capacity and net revenue. *J. Power Sources* **2005**, *144*, 268–279. [CrossRef]

19. Ahmadi, M.; Rastgoo, S.; Mahdavi, Z.; Nasab, M.A.; Zand, M.; Sanjeevikumar, P.; Khan, B. Optimal allocation of EVs parking lots and DG in micro grid using two-stage GA-PSO. *J. Eng.* **2023**, e12237. [CrossRef]
20. Ferraz, R.S.; Ferraz, R.S.; Rueda-Medina, A.C.; Fardin, J.F. Multi-objective Optimization Approach for Allocation of Electric Vehicles Parking Lots and Smart Charging with Distributed Energy Resource. *J. Control. Autom. Electr. Syst.* **2023**, *34*, 1070–1079.
21. Kumar, B.A.; Jyothi, B.; Singh, A.R.; Bajaj, M.; Rathore, R.S.; Tuka, M.B. Hybrid genetic algorithm-simulated annealing based electric vehicle charging station placement for optimizing distribution network resilience. *Sci. Rep.* **2024**, *14*, 7637.
22. Duan, F.; Eslami, M.; Khajehzadeh, M.; Alkhayer, A.G.; Palani, S. An improved meta-heuristic method for optimal optimization of electric parking lots in distribution network. *Sci. Rep.* **2024**, *14*, 20363.
23. Amini, M.H.; Islam, A. Allocation of electric vehicles' parking lots in distribution network. In Proceedings of the ISGT 2014, Washington, DC, USA, 19–22 February 2014;
24. Shaheen, H.I.; Rashed, G.I.; Yang, B.; Yang, J. Optimal electric vehicle charging and discharging scheduling using metaheuristic algorithms: V2G approach for cost reduction and grid support. *J. Energy Storage* **2024**, *90*, 111816. [CrossRef]
25. Ding, X.; Gan, Q.; Shaker, M.P. Optimal management of parking lots as a big data for electric vehicles using internet of things and Long–Short term Memory. *Energy* **2023**, *268*, 126613.
26. Shariatzadeh, M.; Antunes, C.H.; Lopes, M.A. Charging scheduling in a workplace parking lot: Bi-objective optimization approaches through predictive analytics of electric vehicle users' charging behavior. *Sustain. Energy Grids Netw.* **2024**, *39*, 101463. [CrossRef]
27. Shahriar, S.; Al-Ali, A.R.; Osman, A.H.; Dhou, S.; Nijim, M. Machine learning approaches for EV charging behavior: A review. *IEEE Access* **2020**, *8*, 168980–168993.
28. Varone, A.; Heilmann, Z.; Porruvecchio, G.; Romanino, A. Solar parking lot management: An IoT platform for smart charging EV fleets, using real-time data and production forecasts. *Renew. Sustain. Energy Rev.* **2024**, *189*, 113845.
29. Pourvaziri, H.; Sarhadi, H.; Azad, N.; Afshari, H.; Taghavi, M. Planning of electric vehicle charging stations: An integrated deep learning and queueing theory approach. *Transp. Res. Part Logist. Transp. Rev.* **2024**, *186*, 103568.
30. Guo, F.; Xu, H.; Xu, P.; Guo, Z. Design of a reinforcement learning-based intelligent car transfer planning system for parking lots. *Math. Biosci. Eng.* **2024**, *21*, 1058–1081.
31. Adil, M.; Mahmud, M.P.; Kouzani, A.Z.; Khoo, S.Y. Optimal Location and Pricing of Electric Vehicle Charging Stations Using Machine Learning and Stackelberg Game. *IEEE Trans. Ind. Appl.* **2024**, *60*, 4708–4722. [CrossRef]
32. Lv, S.; Xiao, A.; Qin, Y.; Xu, Z.; Wang, X. A decision framework for improving the service quality of charging stations based on online reviews and evolutionary game theory. *Transp. Res. Part Policy Pract.* **2024**, *187*, 104168.
33. Boateng, G.O.; Si, H.; Xia, H.; Guo, X.; Chen, C.; Agyemang, I.O.; Ansari, N. Automated valet parking and charging: A dynamic pricing and reservation-based framework leveraging multi-agent reinforcement learning. *IEEE Trans. Intell. Veh.* **2024**, e12237.
34. Abbasi, M.H.; Arjm Zadeh, Z.; Zhang, J.; Krovi, V.; Xu, B.; Mishra, D.K. A Coupled Game Theory and Lyapunov Optimization Approach to Electric Vehicle Charging At Fast Charging Stations. *IEEE Trans. Veh. Technol.* **2024**, *73*, 14224–14235.
35. Morais, H. New approach for electric vehicles charging management in parking lots considering fairness rules. *Electr. Power Syst. Res.* **2023**, *217*, 109107.
36. Kempton, W. *Vehicle to Grid Power*; FERC: Washington, DC, USA, 2007.
37. Jannati-Oskuee, M.R.; Mojtahedzadeh, S.; Karimi, M. Risk-based joint flexible distribution network expansion planning and allocation of EV parking lots. *Energy Syst.* **2024**, *15*, 1255–1287.
38. Shojaabadi, S.; Abapour, S.; Abapour, M.; Nahavandi, A. Optimal planning of plug-in hybrid electric vehicle charging station in distribution network considering demand response programs and uncertainties. *IET Gener. Transm. Distrib.* **2016**, *10*, 3330–3340.
39. Jeonghwan, J.; Rothrock, L.; Mcdermott, P.L.; Barnes, M. Using the analytic hierarchy process to examine judgment consistency in a complex multiattribute task. *IEEE Trans. Syst. Man, Cybern.-Part Syst. Hum.* **2010**, *40*, 1105–1115.
40. Clement, K.; Haesen, E.; Driesen, J. The impact of charging plug-in hybrid electric vehicles on a residential distribution grid. *IEEE Trans. Power Syst.* **2010**, *25*, 371–380.
41. Sioshansi, R. Modeling the impacts of electricity tariffs on PHEVs. *Oper. Res.* **2012**, *60*, 1–11.
42. Soares, J.; Canizes, B.; Lobo, C.; Vale, Z. Electric vehicle scenario simulator tool for smart grid operators. *Energies* **2012**, *5*, 1881–1899. [CrossRef]
43. Deosaria, T.; Choudhary, S.; Meena, T. Load flow analysis using Forward and Backward sweep, and minimising power losses using Genetic Algorithm. *Int. J. Adv. Eng. Manag. (IJAEM)* **2022**, *4*, 763–772.
44. Zhao, M.; Zhao, H.; Zhao, M. Particle swarm optmization algorithm with adaptive two-population strategy. *IEEE Access* **2023**, *11*, 62242–62260.
45. Khalesi, N.; Rezaei, N.; Haghifam, M.R. DG allocation with application of dynamic programming for loss reduction and reliability improvement. *Int. J. Electr. Power Energy Syst.* **2011**, *33*, 288–295.

46. Fathy, A.; Abdelaziz, Y.A. Competition over resource optimization algorithm for optimal allocating and sizing parking lots in radial distribution network. *J. Clean. Prod.* **2020**, *264*, 121397.
47. Haji-Aghajani, E.; Hasanzadeh, S.; Heydarian-Forushani, E. A novel framework for planning of EV parking lots in distribution networks with high PV penetration. *Electr. Power Syst. Res.* **2023**, *217*, 109156.

algorithms

MDPI

Article

A Multi-Objective Bio-Inspired Optimization for Voice Disorders Detection: A Comparative Study

Maria Habib [1], Victor Vicente-Palacios [2] and Pablo García-Sánchez [1,*]

[1] Department of Computer Engineering, Automation and Robotics, School of Computer and Telecommunication Engineering, University of Granada, 18071 Granada, Spain; mhabib@correo.ugr.es

[2] Department of Statistics, Faculty of Medicine, University of Salamanca, 37008 Salamanca, Spain; victorvicpal@usal.es

* Correspondence: pablogarcia@ugr.es

Abstract: As early detection of voice disorders can significantly improve patients' situation, the automated detection using Artificial Intelligence techniques can be crucial in various applications in this scope. This paper introduces a multi-objective bio-inspired, AI-based optimization approach for the automated detection of voice disorders. Different multi-objective evolutionary algorithms (the Non-dominated Sorting Genetic Algorithm (NSGA-II), Strength Pareto Evolutionary Algorithm (SPEA-II), and the Multi-Objective Evolutionary Algorithm based on Decomposition (MOEA/D)) have been compared to detect voice disorders by optimizing two conflicting objectives: error rate and the number of features. The optimization problem has been formulated as a wrapper-based algorithm for feature selection and multi-objective optimization relying on four machine learning algorithms: K-Nearest Neighbour algorithm (KNN), Random Forest (RF), Multilayer Perceptron (MLP), and Support Vector Machine (SVM). Three publicly available voice disorder datasets have been utilized, and results have been compared based on Inverted-Generational Distance, Hypervolume, spacing, and spread. The results reveal that NSGA-II with the MLP algorithm attained the best convergence and performance. Further, the conformal prediction is leveraged to quantify uncertainty in the feature-selected models, ensuring statistically valid confidence intervals for predictions.

Keywords: voice disorders; K-Nearest Neighbors; MLP; SVM; RF; feature selection; multi-objective optimization; NSGA-II; SPEA-II; MOEA/D

1. Introduction

Early detection of diseases is crucial for medical and healthcare professionals to provide timely interventions with proactive preventive measures. Voice disorders, which affect millions of people around the world, are characterized by abnormal deviations in the voice, observable through changes in pitch, volume, loudness, voice quality, fatigue, and vocal flexibility. These disorders typically manifest through rough vocal quality, breathiness, tremulousness, and pulsed phonation [1]. Artificial intelligence (AI)-driven solutions, therefore, assist voice pathologists and healthcare providers in automatically detecting and diagnosing pathological voices more efficiently and at lower costs [2].

Feature selection is a crucial preprocessing step before developing any machine learning or deep learning model [3]. The presence of irrelevant, redundant, or high-dimensional features can degrade the performance of algorithms. The objective of feature selection is to identify the optimal set of features that enhances algorithm performance. Feature selection

https://doi.org/10.3390/a18060338

techniques can be classified into filter-based, wrapper-based, and embedded-based approaches. Filter-based methods are based on ranking and statistical techniques for feature scoring and selection. On the other hand, wrapper methods utilize a learning algorithm during training to evaluate feature subsets. Lastly, embedded methods, such as genetic programming, learn to predict while training [3]. The challenge with feature selection arises when the number of features is large (i.e., N), thus, the search space becomes exponentially complex, with a size of (2^N). Searching for the optimal feature set in such a vast space becomes computationally expensive, time-consuming, and requires more energy. One of the earlier approaches to address feature selection is the metaheuristic search algorithms. Metaheuristics are stochastic search algorithms incorporating randomization into the search process [4]. Bio-inspired algorithms are metaheuristics and are widely used to solve the feature selection problem [5]. Bioinspired algorithms dynamically search for the best set of features, which enables them to explore various regions in the search space and have more diverse solutions. Bio-inspired algorithms use randomization during the search for the best feature set, allowing exploration of different areas of the search space and leading to more diverse solutions. These algorithms balance exploration and exploitation. Exploration ensures that the algorithm searches globally and avoids getting trapped in local optima, while exploitation focuses on local search to refine solutions.

This paper aims to compare and find the best-performing combination of evolutionary optimizers and machine-learning classifiers for the optimization of feature selection and detection of voice disorders. Thus, the problem of feature selection to improve voice pathology detection is formulated as a multi-objective minimization problem, with two objectives: minimizing the number of features and the error rate. To this end, three well-regarded multi-objective evolutionary algorithms were utilized: NSGA-II [6], SPEA-II [7], and MOEA/D [8]. In addition, four machine learning algorithms were experimented with and served as base learners or internal assessors. The algorithms include the KNN, SVM, MLP, and RF. The experiments were conducted using three datasets, which are the Speech Language Impairment (SLI) [9], Saarbruecken Voice Database (SVD) [10], and the VOice ICar fEDerico II database (VOICED) [11]. Various features were extracted from the voice utterances, including the Mel Frequency Cepstral Coefficients (MFCCs), spectrograms, Linear Predictive Coefficients (LPC), and others [12]. Nonetheless, measuring how ascertain predictive models are in their predictions is important to disclose their reliability in different tasks, such as diagnosing diseases, detecting fraudulent transactions, and others [13]. However, this is not straightforward because models rely on probabilities rather than certainties. One approach to assess certainty is Conformal Prediction (CP). It is used to quantify uncertainty in machine learning models, providing statistically valid confidence estimates across various predictive tasks [14]. By integrating CP, this research improves the robustness and interpretability of feature-selected models, ensuring that predictions in voice disorder detection are reinforced with well-calibrated confidence measures.

The remainder of the paper is organized as follows. Section 2 reviews recent research on multi-objective optimization and voice disorder detection. Section 3 describes the datasets, approach, experimental setup, and evaluation metrics. Section 4 shows the conducted experiments, and discusses the results. Finally, Section 5 is the conclusion and future remarks.

2. Related Works

The detection of voice disorders through machine learning and deep learning techniques has been explored extensively in recent years, as demonstrated in multiple studies. In a notable work by Junior et al. [15], the authors focused on the extraction of handcrafted features, such as energy, entropy, and zero-crossing rate, from the SVD dataset for multi-label detection of voice disorders. They trained several machine and deep learning models, using the Synthetic Minority Oversampling Technique (SMOTE) for oversampling, and found that the convolutional model achieved the highest accuracy of 94.3%. This success highlights the efficacy of machine learning techniques in the automatic detection of voice disorders. Similarly, Rehman et al. [16] utilized several machine learning algorithms, such as decision trees, random forests, and support vector machines (SVM), specifically to detect dysphonia. They worked with both the SVD and Massachusetts Eye and Ear Infirmary (MEEI) datasets, applying filter-based feature selection, and achieved an impressive accuracy of 99.97% with the SVM classifier. These studies underscore the effectiveness of machine learning methods for voice disorder detection. However, the authors in [17] applied Inductive CP with RF to improve the confidence and reliability of silent speech recognition using surface electromyography by providing valid uncertainty estimates with guaranteed error rates. Also, it introduced test-time data augmentation, utilizing unlabeled data to improve prediction performance. This emphasizes the potential of ICP for data augmentation and real-time applications in silent speech recognition.

Further contributions to the field include Yadav et al.'s work [18], where they trained a Sequential Learning Resource Allocation Neural Network (SL-RANN) to detect vocal cord disorders, using discrete Fourier transforms and Mel Frequency Cepstral Coefficients (MFCCs) for preprocessing. The approach achieved a commendable accuracy of 94.35%. Similarly, Albadr et al. [19] used Fast Learning Networks (FLN) in conjunction with MFCC features extracted from the SVD database, achieving a g-mean of 86.81%. Additionally, Sindhu et al. [20] conducted a comprehensive review of deep learning approaches for voice disorder detection, identifying the prevalent use of deep learning for clinical diagnosis and assistive technologies. They also highlighted key gaps in current research, such as the lack of multi-modal approaches, the need to account for variabilities in speech signals (e.g., age, gender, and accent), and the absence of generalizable models that can detect multiple voice disorders simultaneously.

In the realm of optimization for voice disorder detection, several studies have employed evolutionary algorithms. For example, Yildirim et al. [21] proposed an arithmetic optimization algorithm (AOA) for detecting Parkinson's disease (PD), using feature maps extracted by eight convolutional neural network (CNN) architectures from spectrogram images. This approach, combined with machine learning algorithms like KNN and SVM, achieved an average accuracy of 96.62%. Similarly, Singh et al. [22] introduced a whale-ant (WO-ANT) optimization algorithm for PD detection using speech samples, which achieved 90.29% accuracy. Akila et al. [23] further explored PD detection using a neural network optimized with a multi-agent salp swarm (MASS) algorithm, obtaining an F1-score of 0.995. However, it is important to note that these studies were focused solely on Parkinson's disease. Also, authors in [24] enhanced the detection of Parkinson's disease by using the Particle Swarm Optimization to assign weights for the acoustic features based on their relevance to the classifiers. Using the Oxford PD dataset and 5-fold cross-validation, the method improved the classification performance, where KNN achieved an accuracy of 97.44% and sensitivity of 98.02%.

Evolutionary multi-objective optimization (EMO) techniques have also been successfully applied in various domains, including voice disorder detection. Nasrolahzadeh et al. [25] utilized the NSGA-II algorithm to select optimal wavelet features from spontaneous speech for early Alzheimer's disease detection, achieving 98.33% accuracy. Additionally, Altay et al. [26] applied a multi-objective evolutionary association-rule mining approach to acoustic voice characteristics for diagnosing PD, where the NICGAR algorithm performed the best among the methods tested. Sheth et al. [27] proposed a multi-objective Jaya algorithm for PD detection, formulating the problem as a weighted sum fitness function for feature reduction and accuracy maximization, achieving 92.26% accuracy with a KNN classifier. Finally, Patra et al. [28] demonstrated the effectiveness of MOEA/D for optimized feature selection in medical diagnosis.

Recently, despite the widespread use of machine learning methods in various fields, and specifically in voice disorder detection, there are some limitations. For instance, many research studies used machine and deep learning algorithms to detect pathological voices, a high proportion of them targeted Parkinson's patients, disregarding other diseases. Nonetheless, many of the studies which used evolutionary optimizers formulated the optimization problem as a single-objective optimization problem to reduce the high dimensionality of the problem. Very few research studies examined evolutionary multi-objective optimization for pathological voice detection, where they emphasized Parkinson's disease. This study attempts to address the limitation by employing multi-objective evolutionary optimization with various voice disorders that eventually reduce the number of features and power consumption. Table 1 concludes related studies for voice disorder detection. As can be noticed in the table, most of the studies rely on accuracy to evaluate proposed methods. In this study, the generational distance and hypervolume metrics are employed solely to evaluate the effectiveness of optimization in the context of multi-objective evolutionary algorithms, and not to compare the classification performances among different models.

Table 1. Summary of related works on voice disorder detection.

Paper	Method	Results	Objective
Junior et al. [15]	Hand-crafted features + CNN + SMOTE	Accuracy: 94.3%	Multi-label voice disorder detection using SVD dataset
Rehman et al. [16]	ML algorithms (DT, RF, SVM) + filter-based feature selection	Accuracy: 99.97% (SVM)	Dysphonia detection using SVD and MEEI datasets
Zhang et al. [17]	Inductive CP + RF + Test-time augmentation	Reliable uncertainty estimates	Improve confidence in silent speech recognition
Yadav et al. [18]	SL-RANN + DFT + MFCC	Accuracy: 94.35%	Vocal cord disorder detection
Albadr et al. [19]	Fast Learning Networks (FLN) + MFCC (SVD)	G-mean: 86.81%	Voice disorder detection using FLN
Sindhu et al. [20]	Literature Review	–	Identified research gaps in deep learning for voice disorder detection
Yildirim et al. [21]	AOA + CNN features + KNN/SVM	Accuracy: 96.62%	Parkinson's disease detection using spectrograms
Singh et al. [22]	Whale-Ant (WO-ANT) optimization	Accuracy: 90.29%	PD detection using speech samples
Akila et al. [23]	NN + Multi-agent Salp Swarm (MASS)	F1-score: 0.995	PD detection using optimized neural network
Nasrolahzadeh et al. [25]	NSGA-II + wavelet features	Accuracy: 98.33%	Alzheimer's detection from spontaneous speech
Altay et al. [26]	NICGAR (multi-objective association-rule mining)	–	Diagnose PD using acoustic characteristics
Sheth et al. [27]	Multi-objective Jaya + KNN	Accuracy: 92.26%	PD detection with feature reduction
Patra et al. [28]	MOEA/D	–	Feature selection for medical diagnosis

3. Material and Methods

This section describes the evolutionary multi-objective approach for voice disorder detection. It starts by explaining the datasets and extracted features, how the problem is formulated and the methods used. Finally, the evaluation measures and the experimental setups are described.

3.1. Datasets

The datasets utilized in this study encompass a range of speech recordings from various sources, each designed to capture distinct voice characteristics and pathologies, providing a comprehensive foundation for analysing and detecting voice disorders. The SLI dataset was developed by the Czech Technical University and contains recordings from 15 boys and 35 girls. Their ages range from 4 to 12 years. All recordings are in WAV format, featuring a 44.1 kHz sampling rate and a 16-bit depth. Entirely, the dataset includes 3258 recordings, with 2173 from patients and 1083 from healthy individuals [9]. In addition, the SVD dataset was downloaded from [10], which comprises recordings from over 2000 individuals, including voice utterances from 1354 individuals with more than 71 different pathologies, alongside 687 healthy individuals. The duration of these voice recordings ranges from 1 to 3 s, and four specific voice disorders were selected from the database: Hyperfunctional Dysphonia, Hypofunctional Dysphonia, Dysphonia, and Laryngitis. These disorders did not cause significant degradation in voice utterances, so they are selected as the initial scope of this study. Ultimately, the SVD dataset contains 1079 voice samples, of which 446 are pathological, consisting of vowel recordings of /a/, /i/, and /u/, pronounced in three different intonations: high, low, and neutral. Finally, the VOICED dataset consists of recordings of subjects holding the vowel 'a' for five seconds continuously. The VOICED dataset comprises 208 recordings sampled at 8000 Hz with a 32-bit resolution. It is divided between 58 healthy samples and 150 with vocal fold disorders [11].

To analyze these datasets, key features from raw speech signals were extracted using the Librosa library [29]. Several spectral and rhythm features are extracted, including Mel-Frequency Cepstral Coefficients (MFCC), Mel spectrogram, frequency and autocorrelation Tempogram, Chromagram or the power spectrum, spectral contrast, roll-off frequency, and flatness. As well as, tonal centroid features, spectrogram poly-feature, zero-crossing rate, spectral centroid and bandwidth, and the root mean square error of each frame. All features are computed per frame, and the mean and standard deviation are used to create the dataset. The final created dataset contains 849 features, and one binary class of two labels (healthy or pathological). The strength of extracted features is that the MFCC encodes spectral envelopes through mathematical transformations, while spectrograms visualize frequency changes over time, aligning better with human auditory perception. Furthermore, the Tempogram analyzes rhythmic patterns, and chroma features represent tonal content by grouping frequencies into twelve pitch classes. Other spectral features, including spectral centroid, bandwidth, flux, flatness, and zero-crossing rate, provide further insights into the signal's frequency content and waveform characteristics.

3.2. Proposed Method

This section describes the proposed method, starting with the algorithms compared, the individuals encoding, fitness function, evaluation measures, and the experimental setup.

3.2.1. Algorithms Compared

Multi-objective problems have two or three simultaneous conflicting objectives. Although improving one objective might lead to a degradation in the other one, multi-objective optimization algorithms search to have the optimal set of solutions that are called the Pareto set solutions. The Pareto solutions are trade-off solutions where no solution is better than the others. Mathematically, a multi-objective optimization problem with constraints is defined by Equations (1)–(3) [30]. In Equation (1), $F(X)$ is a minimization function. X is an n-dimensional vector of decision variables, which is defined by $X = x_1, \ldots, x_n$.

The k variable is the number of objective functions. Equations (2) and (3) present the inequality ($g(x)$) and equality ($h(x)$) constraints, respectively.

$$Minimise\ F(X) = [f_1(X),\ f_2(X), \ldots, f_k(X)] \tag{1}$$

$$g_m(x) \leq 0,\ m = 1,\ 2, \ldots M \tag{2}$$

$$h_j(x) = 0,\ j = 1,\ 2, \ldots J \tag{3}$$

Comparing the obtained solutions of multi-objective optimization is done via the Pareto-Dominance. Consider a two-objective minimization problem with two solutions, represented as $\vec{x_1}$ and $\vec{x_2}$. We say that $\vec{x_1}$ dominates $\vec{x_2}$, denoted as ($\vec{x_1} \preceq \vec{x_2}$), if the condition described in (Equation (4)) is fulfilled.

$$\forall i : f_i(\vec{x_1}) \leq f_i(\vec{x_2}) \quad \& \quad \exists j : f_j(\vec{x_1}) < f_j(\vec{x_2}),\ where\ i \,\&\, j \in \{1, 2, \ldots k\} \tag{4}$$

This means that all elements of $f(\vec{x_1})$ are no worse than their corresponding elements in $f(\vec{x_2})$, with at least one element being strictly better. Figure 1 illustrates the concept of dominance, highlighting the green and yellow colored solutions. $\vec{x_b}$ is superior in objective (1) but inferior in objective (2), while $\vec{x_a}$ excels in objective (2) and performs worse in objective (1). Consequently, both $\vec{x_a}$ and $\vec{x_b}$ are considered non-dominated solutions, forming part of the Pareto optimal set, denoted as $P^* = \vec{x^*} \in \Omega$. In contrast, solution c is dominated by both solutions a and b, which are superior.

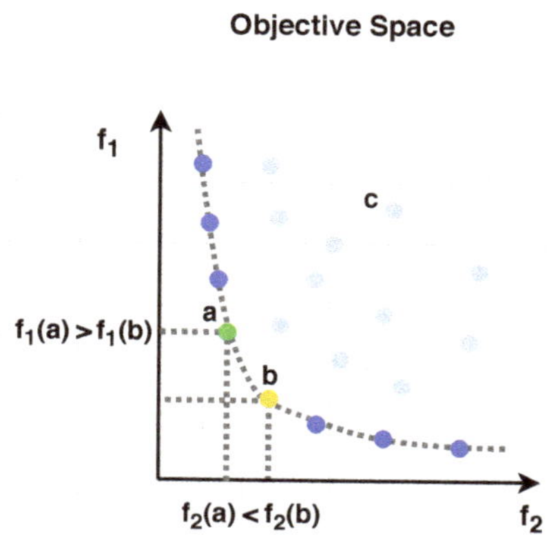

Figure 1. A depiction of dominance, where both f_1 and f_2 are minimization functions. The grey points represent dominated solutions, while the blue dots indicate non-dominated solutions.

Four well-regarded evolutionary multi-objective optimization algorithms were used and compared for the optimization of voice disorder detection (The NSGA-II, SPEA-II, and MOEA/D). The NSGA-II is a modified version of the NSGA algorithm and improves over the first version by being faster and incorporating a better sorting and selection strategy. Its objective is to find a set of solutions that are non-dominated with respect to each other. The NSGA-II algorithm relies on the genetic algorithm while employing non-dominated sorting and Crowding-distance fitness. It works by producing an initial population of solutions and then evaluating them based on multiple objectives. It ranks solutions into different Pareto-fronts using non-dominated sorting, where solutions in the first front are non-dominated by any others. To preserve diversity, it calculates a crowding distance, ensuring solutions are spread across the objective space. Through selection, crossover, and mutation, NSGA-II creates new generations, combining parent and offspring populations to maintain the best solutions. This process is repeated until a stopping criterion is met, ultimately yielding a set of Pareto-optimal solutions [6].

The SPEA-II is an advanced multi-objective optimization algorithm developed to find and maintain a diverse set of Pareto-optimal solutions. It works by keeping an external archive of non-dominated solutions, allocating fitness values based on both the number of solutions a candidate dominates (strength) and a density estimation to ensure diversity. At each generation, individuals are selected based on their fitness, and genetic operators like crossover and mutation are applied to generate new solutions. The archive is updated with the best solutions, gradually guiding the algorithm towards a well-distributed Pareto-front [7].

The MOEA/D is an optimization algorithm designed to tackle complex multi-objective problems by breaking them into simpler subproblems. Each subproblem corresponds to a unique weight vector representing a trade-off between objectives. The algorithm evolves and improves solutions by solving these subproblems concurrently, using genetic operators such as crossover and mutation. Solutions to each subproblem are improved with the help of adjacent subproblems, ensuring collaboration and diversity. By dividing the overall problem into smaller, more manageable parts, MOEA/D efficiently balances exploration and exploitation, driving the population toward a well-distributed set of Pareto-optimal solutions [8].

3.2.2. Individual Encoding

In this study, the detection problem of voice disorders is formulated as a multi-objective optimization problem, utilizing NSGA-II, SPEA-II, and MOEA/D. The objective is twofold: to minimise the number of features used for classification while simultaneously minimizing the error rate of the predictive model. The datasets employed in this work include the SLI, SVD, and VOICED datasets. Each candidate solution represents a possible subset of features in the evolutionary optimization framework. The solutions are encoded as binary vectors, where each bit corresponds to a feature. A value of '1' refers to the feature being selected, while '0' means the feature is excluded. In other words, if a dataset has 849 features, a binary string of length 849 represents one possible feature subset, where the activated (1) and deactivated (0) features indicate which features are selected for the classification process. Figure 2 illustrates a solution.

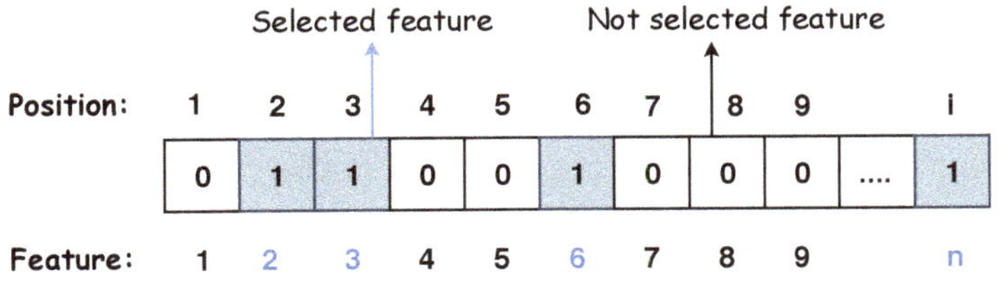

Figure 2. A representation of a solution, where the number of features is n, and i is the index.

3.2.3. Fitness Function

The problem is addressed by simultaneously optimizing two conflicting objectives (i.e., reducing the number of selected features and minimizing the classification error rate). Equation (5) presents the fitness equation, where l is the number of selected features, and Acc is the classification accuracy. The first objective function $f_1(x)$ points for the number of selected features. Whereas, the second objective function $f_2(x)$ corresponds to the classification error rate, which is computed as (1 − accuracy).

$$F(x) = \begin{cases} f_1(x) = l & , l \in \mathbb{R}^+ \\ f_2(x) = 1 - accuracy & , accuracy \in \mathbb{R}^+ \end{cases} \quad (5)$$

The three optimization algorithms produce a Pareto-front of non-dominated solutions, which provides a set of trade-off solutions. During the evolutionary process, those solutions are evaluated using four machine learning classifiers (RF, KNN, MLP, and SVM). Each solution (i.e., feature subset) is passed through these classifiers, and its performance is assessed based on the error rate and the number of selected features. The evaluation process involves cross-validation evaluation of the datasets, ensuring that the selected feature subsets generalize well across different training and testing splits. Figure 3 presents the methodology.

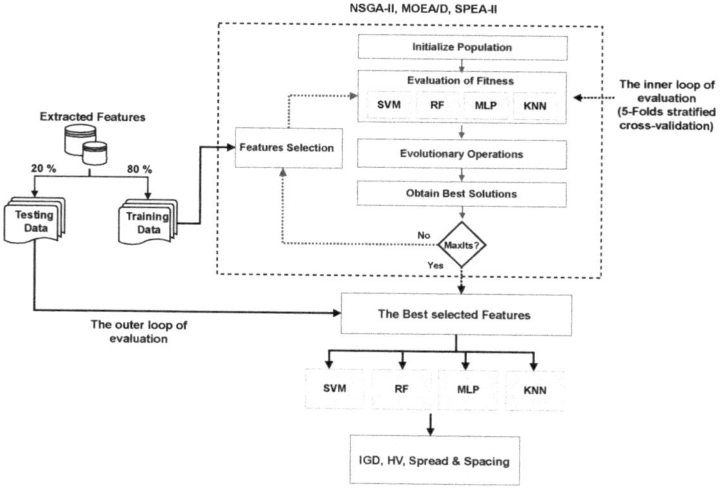

Figure 3. Architectural design of the methodology.

3.3. Evaluation Measures

This section presents the evaluation metrics used to assess the multi-objective optimized model depending on IGD, Hypervolume, and spread and spacing.

3.3.1. IGD

The Inverted Generational Distance (IGD) is a modified generational distance (GD) metric. The GD measures the distances from each solution vector (from the approximation front) to the closest true Pareto-front solution. Whereas, the IGD computes the opposite, the distances from each solution in the true Pareto-front to its closest solution in the approximation front, as given by Equation (6). However, because the true Pareto-front is not known, the IGD was computed after the execution of all solutions across all runs per dataset. A smaller value of IGD indicates a better solution [31].

$$IGD_p(A, R) = \frac{1}{R} \times \left(\sum_{r \in R} min\ d(r, a)^p \right)^{1/p} \quad (6)$$

3.3.2. Hypervolume

The Hypervolume metric (HV) indicates the quality of non-dominated estimated solutions. It calculates the area enclosed by the estimated Pareto-front and the true reference Pareto solution.

3.3.3. Spread & Spacing

Spread and spacing measure how well the estimated solutions are evenly distributed over the Pareto-front and not clustered in certain regions. A higher spread value is a better indicator of the estimated solutions' quality. The spacing refers to how much the distances of the adjacent estimated solutions are consistent and uniform. A lower value indicates that solutions are more evenly spread.

3.4. Uncertainty Quantification

To quantify the uncertainty associated with machine learning models, conformal prediction (CP) is adopted. CP is a statistical framework that measures the uncertainty in predictions of any predictive model, from regression to deep learning. It works by: first, training a model; second, computing conformity scores on a calibration set; and third, by constructing prediction sets for new data that guarantee a user-defined confidence level [14].

Using a held-out calibration set, nonconformity scores are computed as shown in Equation (7), where $p(y_i^{\text{true}})$ represents the predicted probability of the model for the true class y_i, and s_i represents the nonconformity score. These scores calibrate prediction sets $\mathcal{C}(X_{\text{test}})$ to achieve marginal coverage guarantees, as shown by Equation (8), at a significance level α (given that $\alpha = 0.05$ for coverage 95%) [14].

$$s_i = 1 - p(y_i^{\text{true}}) \tag{7}$$

$$\mathbb{P}\big(y_{\text{test}} \in \mathcal{C}(X_{\text{test}})\big) \geq 1 - \alpha \tag{8}$$

3.5. Experimental Setup

The experimental setup for optimizing the detection of voice disorders using NSGA-II, SPEA-II, and MOEA/D was conducted using the PyGMO library [32], applying default parameters for all the algorithms and machine learning models involved. The machine learning algorithms are implemented using the Scikit-learn library [33]. The default parameters are presented in Table 2.

Table 2. Default values of KNN, RF, MLP, and SVM in Scikit-learn.

Algorithm	Parameter	Default Value
KNN	(n_neighbors)	5
	Weights	Uniform
	Algorithm	Auto
	Metric	Minkowski
RF	n_estimators	100
	Criterion	Gini
	max_depth	None
	min_samples_split	2
	min_samples_leaf	1
	Bootstrap	True

Table 2. *Cont.*

Algorithm	Parameter	Default Value
MLP	hidden_layer_sizes	(100,)
	Activation	Relu
	Solver	Adam
	Alpha	0.0001
	Learning Rate	Constant
	max_iter	200
	Tolerance (tol)	1×10^{-4}
SVM	Regularization Parameter (C)	See Table 3
	Kernel	RBF
	Degree	3
	Gamma	See Table 3
	Coef0	0.0
	Tolerance (tol)	1×10^{-3}

Regarding the SVM algorithm, several experiments were conducted to find the best values of the γ and C hyper-parameters per dataset [34]. In this regard, cost and gamma for each dataset are set as in Table 3.

Table 3. Best SVM parameters (Cost and Gamma) for the SLI, SVD, and VOICED datasets.

Dataset	Cost (C)	Gamma (γ)
SLI	33.20983613	9.700×10^{-6}
SVD	5.37391496	1.880×10^{-6}
VOICED	16.89381839	4.590×10^{-5}

Furthermore, Table 4 summarizes the used hyper-parameters of NSGA-II, SPEA-II, and MOEA/D as the default values defined in PyGMO library. Those hyper-parameters control key aspects of the optimization behaviour, such as genetic operators, neighborhood structures. Regarding the MOEA/D, the neighborhood-based selection and aggregation function (Tchebycheff) are critical to its decomposition strategy.

Table 4. Default hyper-parameters for multi-objective algorithms in PyGMO.

Parameter	NSGA-II	SPEA2	MOEA/D
Population Size	30	30	30
Crossover Prob.	0.95	0.95	1.0
Mutation Prob.	0.01	0.01	-
Distribution Index (SBX)	10	10	20
Distribution Index (Mutation)	10	50	20
Number of Generations	50	50	50
Neighborhood Size	N/A	N/A	20
Decomposition Method	N/A	N/A	Tchebycheff

The performance evaluation was performed via five-fold cross-validation of the training subset to ensure robust performance assessment. The population and offspring sizes were set to 30, with 50 generations per run. To account for variability, each experiment was repeated across 15 independent runs. The HV was computed using the worst-performing solution among all Pareto-fronts, and the reference point for the IGD was derived from all solutions across all runs per dataset. The experiments were executed on a dedicated server to handle the computational demands of the process. The server has a processor of 12th Gen Intel(R) Core(TM) i9-12900KF, and a memory of 125 GB.

4. Results and Discussion

This section presents the experiments conducted and the results obtained for NSGA-II, MOEA-D, and SPEA-II on three datasets (SLI, SVD, and VOICED). The three optimizers were assessed using four base machine learning classifiers (RF, KNN, MLP, and SVM).

Table 5 presents the performance of three multi-objective optimization algorithms —NSGA-II, MOEA/D, and SPEA-II— when paired with the KNN classifier across three datasets: SLI, SVD, and VOICED. It provides the average (AVG) and standard deviation (STD) for four key metrics: Inverted Generational Distance (IGD), Hypervolume (HV), Spacing, and Spread. Analyzing the IGD metric, NSGA-II achieved the lowest IGD value on the SLI dataset (15.506), indicating that it closely approximates the Pareto-front in this case. On the other hand, SPEA-II performed best on the SVD and VOICED datasets, with IGD values of 31.787 and 30.161, respectively. This suggests that SPEA-II is effective at finding solutions near the Pareto-front in these datasets.

In terms of HV, which measures the volume covered by the solutions in objective space, both NSGA-II and SPEA-II obtained the highest values across all datasets. Specifically, NSGA-II performed best on the SLI dataset (20.139), while SPEA-II excelled on the SVD (15.245) and VOICED (21.192) datasets.

Regarding the Spacing and Spread metrics, which evaluate solution distribution and diversity, SPEA-II generally outperformed in Spacing, with values of 0.900 and 0.765 for the SLI and SVD datasets. This indicates a more evenly distributed set of solutions. Conversely, NSGA-II achieved lower values for Spread on the SLI dataset, signifying a balanced diversity across objectives.

Overall, NSGA-II performs robustly on the SLI dataset, while SPEA-II demonstrates strong performance on the SVD and VOICED datasets, highlighting the dataset-specific strengths of the different optimizers.

Table 5. Optimization results of NSGA-II, MOEA/D, and SPEA-II with KNN classifier across three datasets (SLI, SVD, and VOICED). The best results were highlighted with boldface style.

Algorithm	Optimizer	Datasets	IGD		HV		Spacing		Spread	
			AVG	STD	AVG	STD	AVG	STD	AVG	STD
KNN	MOEA/D	SLI	95.244	9.578	5.117	1.078	1.750	1.848	**1.724**	0.437
		SVD	95.645	6.450	3.577	1.032	1.108	1.339	1.593	0.484
		VOICED	95.912	11.388	5.158	1.493	1.376	1.841	**1.709**	0.428
	SPEA-II	SLI	22.735	11.863	19.865	3.070	**0.900**	0.691	1.623	0.304
		SVD	**31.787**	11.190	**15.245**	2.013	0.765	0.583	**1.844**	0.128
		VOICED	**30.161**	11.881	**21.192**	2.285	1.243	1.308	1.575	0.385
	NSGA-II	SLI	**15.506**	9.425	**20.139**	2.595	1.163	0.681	1.589	0.209
		SVD	36.502	12.645	13.836	1.711	**0.528**	0.599	1.728	0.356
		VOICED	34.781	15.437	19.868	2.881	**1.073**	1.288	1.654	0.356

Table 6 displays the performance results of the RF classifier. Looking at IGD, SPEA-II achieves the lowest values on SLI and VOICED, suggesting more precise Pareto-front approximations, while NSGA-II performs similarly on SVD, with an IGD of 31.781. In the HV metric, SPEA-II consistently scores highest across all datasets, indicating a strong coverage in objective space. NSGA-II follows closely, particularly with VOICED where it achieves a high HV of 10.585. Spacing and Spread metrics indicate solution distribution and diversity. MOEA/D's spacing is minimal on SVD and VOICED, with values near zero (0.032 and 0.031), implying evenly distributed solutions. In contrast, Spread is generally consistent across optimizers. Overall, SPEA-II demonstrates reliable performance across datasets with balanced spacing and spread.

Table 6. Optimization results of NSGA-II, MOEA/D, and SPEA-II with RF classifier across three datasets (SLI, SVD, and VOICED). The best results were highlighted with boldface style.

Algorithm	Optimizer	Datasets	IGD		HV		Spacing		Spread	
			AVG	STD	AVG	STD	AVG	STD	AVG	STD
RF	MOEA/D	SLI	96.778	8.740	2.127	0.601	1.531	1.617	**1.851**	0.334
		SVD	99.845	12.100	2.735	0.499	**0.032**	0.121	1.065	0.242
		VOICED	102.911	11.032	5.645	0.977	**0.031**	0.091	1.134	0.329
	SPEA-II	SLI	**24.067**	16.246	**6.658**	1.814	0.806	0.943	1.812	0.303
		SVD	35.683	11.060	5.145	0.663	0.443	0.527	**1.716**	0.377
		VOICED	35.017	9.462	10.327	1.392	0.265	0.173	1.773	0.319
	NSGA-II	SLI	32.548	10.323	5.810	0.550	**0.361**	0.240	1.771	0.388
		SVD	**31.781**	6.828	**5.585**	0.615	0.217	0.190	1.706	0.426
		VOICED	**31.795**	11.075	**10.585**	1.300	0.541	0.509	**1.847**	0.276

Table 7 presents the results of the comparison between the three optimizers with the MLP classifier across three datasets. SPEA-II generally performed well, achieving the lowest IGD for SLI and SVD, suggesting strong proximity to the true Pareto-front, while NSGA-II had the lowest IGD for VOICED. For HV, SPEA-II consistently achieved the highest averages, especially on SVD and VOICED, indicating good coverage of the objective space. SPEA-II also excelled in Spacing, achieving the most uniform solution distribution, especially notable in the VOICED dataset. Spread results showed that both SPEA-II and NSGA-II maintained good solution diversity, with SPEA-II achieving the best values on SVD. Overall, SPEA-II shows strong and consistent optimization performance across metrics and datasets, suggesting it is the most robust algorithm for MLP-based multi-objective optimization.

Table 7. Optimization results of NSGA-II, MOEA/D, and SPEA-II with MLP classifier across three datasets (SLI, SVD, and VOICED). The best results were highlighted with boldface style.

Algorithm	Optimizer	Datasets	IGD		HV		Spacing		Spread	
			AVG	STD	AVG	STD	AVG	STD	AVG	STD
MLP	MOEA/D	SLI	99.378	12.424	1.455	0.462	0.802	1.564	1.298	0.447
		SVD	97.978	12.900	4.019	0.825	1.444	2.072	1.530	0.491
		VOICED	96.312	13.059	8.459	1.769	1.556	3.275	1.266	0.441
	SPEA-II	SLI	**19.096**	9.635	3.880	0.432	**0.659**	0.855	**1.685**	0.368
		SVD	**23.348**	9.737	**14.340**	2.088	**0.689**	0.813	**1.858**	0.241
		VOICED	31.076	11.602	**23.985**	4.138	**0.269**	0.338	1.593	0.421
	NSGA-II	SLI	22.941	10.620	**4.049**	0.668	1.060	1.617	1.676	0.549
		SVD	25.337	9.551	14.044	1.372	0.712	0.429	1.820	0.298
		VOICED	**30.488**	13.773	22.070	2.879	1.048	1.891	**1.765**	0.310

Table 8 displays the optimization performance of NSGA-II, MOEA/D, and SPEA-II with the SVM classifier across (SLI, SVD, and VOICED). NSGA-II achieves the best IGD on SLI and VOICED, indicating the highest proximity to the true Pareto-front, while SPEA-II performs best on SVD. For HV, NSGA-II outperforms the other algorithms on SLI, and SPEA-II achieves a high HV on SVD, implying superior objective space coverage on those datasets. In terms of Spacing, SPEA-II performs well with the lowest variance across SLI and SVD, indicating a more uniform distribution. MOEA/D shows mixed performance, and all algorithms report zero values for all metrics on VOICED, likely due to convergence issues. Overall, NSGA-II and SPEA-II demonstrate strong performance across the metrics, particularly on SLI and SVD, with NSGA-II generally excelling in HV and IGD.

Table 8. Optimization results of NSGA-II, MOEA/D, and SPEA-II with SVM classifier across three datasets (SLI, SVD, and VOICED). The best results were highlighted with a boldface style.

Algorithm	Optimizer	Datasets	IGD		HV		Spacing		Spread	
			AVG	STD	AVG	STD	AVG	STD	AVG	STD
SVM	MOEA/D	SLI	99.978	13.368	3.092	0.967	1.477	1.747	**1.920**	0.246
		SVD	104.645	8.046	1.644	0.462	1.472	1.461	**1.912**	0.246
		VOICED	90.378	10.354	0.000	0.000	0.000	0.000	0.000	0.000
	SPEA-II	SLI	23.230	8.986	8.035	1.178	**0.726**	0.990	1.297	0.574
		SVD	**30.228**	12.511	**7.797**	1.090	**1.369**	1.120	1.771	0.169
		VOICED	72.978	9.934	0.000	0.000	0.000	0.000	0.000	0.000
	NSGA-II	SLI	**22.000**	9.565	**9.392**	1.450	1.327	1.060	1.796	0.141
		SVD	35.040	8.447	1.766	0.181	1.950	1.190	1.809	0.200
		VOICED	**24.059**	12.257	0.000	0.000	0.000	0.000	0.000	0.000

Figure 4 depicts the approximated Pareto-front for the three datasets from all optimizers and all classifiers. Figure 4A–C present the best obtained Pareto-front from the NSGA-II for SVD, VOICED, and SLI, respectively, among all classifiers. Figure 4D shows the actual Pareto-front (the non-dominated solutions) obtained by all classifiers over all runs.

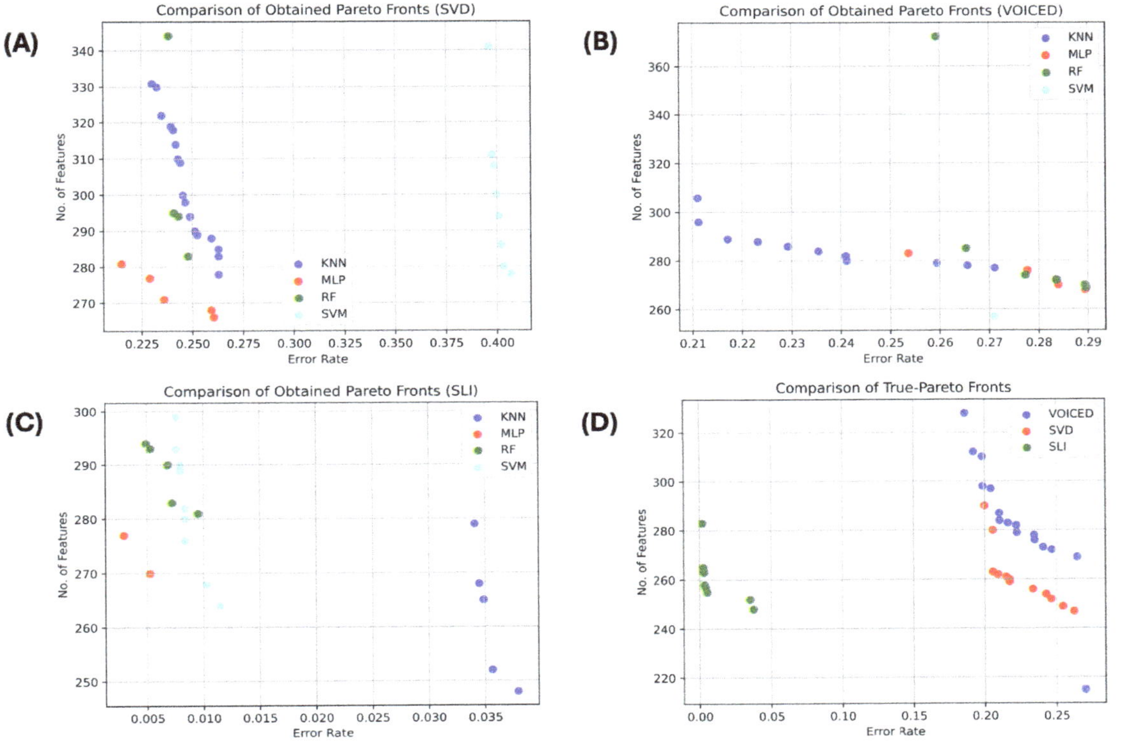

Figure 4. Comparison of the best Pareto-Front per dataset and overall datasets.

Table 9 shows the required optimization time (i.e., training time) of the NSGA-II optimizer across KNN, MLP, RF, and SVM applied over the three datasets, with the average and standard deviation values computed. The optimization time represents the needed time for the optimization phase of the evolutionary algorithm to minimize the number of features and error rate. The results indicate significant variations in optimization time depending on the classifier and dataset. The MLP and RF classifiers are the most computationally intensive, particularly on the SLI dataset, with average times of 7619.2 s and 7936.3 s, respectively, which may reflect the higher complexity and resource demand of these models in training with NSGA-II. For smaller datasets like VOICED, all classifiers have lower optimization times, with KNN and SVM exhibiting particularly fast runtimes (8.319 s for KNN and 14.633 s for SVM). SVM shows consistently low optimization times across datasets, which may make it more practical for large-scale or real-time applications requiring NSGA-II optimization. The table highlights the computational demands of NSGA-II optimization, which vary substantially across classifiers and datasets, impacting its feasibility depending on the available resources and time constraints.

Table 9. Optimization time of the NSGA-II over all classifiers and datasets.

Algorithm	Datasets	Optimization Time (s)	
		AVG	STD
KNN	SLI	61.408	1.051
	SVD	20.021	0.266
	VOICED	8.319	0.947
MLP	SLI	7619.222	913.322
	SVD	3237.941	450.743
	VOICED	849.334	192.432
RF	SLI	7936.300	197.448
	SVD	4493.705	580.553
	VOICED	1045.188	65.990
SVM	SLI	834.729	25.778
	SVD	242.100	6.263
	VOICED	14.633	0.130

Furthermore, a statistical analysis is conducted to guarantee the significant difference in performance between the three evolutionary optimizers across the four machine learning algorithms and datasets. First, the Kolmogorov–Smirnov (KS) normality test is used to ensure the validity of the data distribution before computing the Analysis of Variance (ANOVA) test. If at least one dataset was abnormal, the Kruskal–Wallis test is calculated instead. The algorithms' performance was evaluated using IGD and Hypervolume metrics over 15 independent runs. The test results exhibited a low p-value less than the significance level of 0.05, indicating a significant difference in the performance of the algorithms.

Generally speaking, NSGA-II is the best optimizer overall. While KNN performs best for SLI, MLP outperforms others in SVD, and SVM achieves the best IGD for VOICED, with MLP leading in HV, making NSGA-II and MLP a strong but not always the optimal solution. NSGA-II and MLP were used to compute the non-conformity scores for the three datasets. The split conformal prediction with softmax-based nonconformity scores is employed. Using the MAPIE Library [35], the CPs for all runs and all Pareto-fronts were computed, with the average (Avg.) and standard deviation (Std.) were reported for the non-conformity scores, as shown in Table 10. The results show a strong performance of the SLI dataset with a high accuracy of 0.987 and low nonconformity scores (0.015 ± 0.079), indicating well-calibrated predictions and reliable uncertainty estimates. Conversely, the SVD and VOICED datasets exhibit lower accuracy (0.743 and 0.706, respectively) and higher nonconformity scores (0.253–0.318), suggesting increased model uncertainty or data challenges.

Table 10. Conformal predictions of NSGA-II and MLP.

Dataset	Accuracy	Nonconformity (Avg.)	Nonconformity (Std.)
SLI	0.987	0.015	0.079
SVD	0.743	0.253	0.364
VOICED	0.706	0.318	0.350

Table 11 presents a comparative overview of our proposed methods along with recent State-of-the-Art approaches from the literature. Although we acknowledge that the comparison is not strictly fair due to differences in datasets, feature extraction strategies, and optimization objectives, the table is intended to provide a contextual reference to understand current trends.

In particular, while many previous studies report high performance—particularly on the full SVD or mixed datasets (e.g., SVD + MEEI)—our experiments were conducted on different bases. Specifically, we used a subset of the SVD dataset, employed distinct handcrafted features (849 in total), and focused on underexplored datasets such as VOICED and SLI, for which, to the best of our knowledge, no multi-objective evolutionary optimization approaches have been previously applied.

The approach designed in this study emphasizes a multi-objective formulation using algorithms like NSGA-II to jointly minimize classification error and reduce the feature set. Despite relatively lower classification metrics in some cases, our methods achieve a threefold reduction in dimensionality, improving computational efficiency and interpretability.

Overall, although the table does not represent a direct benchmarking effort, it reinforces the unique value of this study's contributions. Specifically, it emphasizes the integration of multi-objective optimization and the exploration of lesser-studied datasets. Thus, adding diversity and new directions to the existing body of research.

Table 11. Comparison of Proposed Method with State-of-the-Art Approaches.

Study	Method	Dataset	Accuracy	F1-Score	Precision	Recall
Junior et al. [15]	CNN + SMOTE	SVD	>90%	94.3%	-	-
Rehman et al. [16]	SVM + Feature Sel.	SVD, MEEI	99.97%	99%	-	-
Yadav et al. [18]	SL-RANN	SVD	94.35%	-	-	-
Albadr et al. [19]	FLN + MFCC	SVD	84.64%	86.80%	97.39%	86.05%
Yildirim et al. [21]	AOA + CNN + SVM	PD	96.62%	-	-	-
Akila et al. [23]	MASS-NN	PD (UCI)	99.1%	99.5%	-	-
Proposed Method	NSGA-II-based SVM	VOICED	69.0%	81.2%	71.8%	93.3%
Proposed Method	NSGA-II-based MLP	SVD	63.9%	66.7%	53.8%	87.6%
Proposed Method	NSGA-II-based KNN	SLI	81.3%	86.1%	85.5%	86.7%

5. Conclusions

This comparative research study presents a new multi-objective and bio-inspired machine learning optimization approach for improving the accuracy of voice disorder detection. Three datasets were utilized along with three multi-objective optimizers (NSGA-II, SPEA-II, and MOEA/D) and four base classifiers (RF, KNN, MLP, and SVM). In this study, the problem is formulated as a multi-objective feature selection, which aims to minimize the number of features and the error rate. Several spectral and time-domain features were extracted from SLI, SVD, and VOICED datasets and labelled as normal and abnormal. Comparing algorithms based on the IGD, HV, Spread, and Spacing metrics showed that the three optimizers exhibited competitive performance, while the NSGA-II algorithm disclosed a powerful capability for identifying pathological voices. Also, the best classifiers per dataset are KNN for SLI (best IGD & HV), MLP for SVD (best IGD), and SVM for VOICED (best IGD), with MLP also excelling in HV. Although NSGA-II with MLP is not always the best, it remains a strong choice. Furthermore, by integrating conformal prediction, this study ensures statistically feasible uncertainty quantification in feature-selected models, enhancing the reliability of voice disorder detection even in high-dimensional settings.

Future research will investigate optimizing parameters and refining feature selection more closely, as these are important steps for promoting the performance of speech disorder detection. Incorporating advanced algorithms, such as deep neural networks, can further improve performance. Additionally, recent studies have shown the effectiveness of large language models (LLMs) in automatic speech recognition and speech analysis. Thus, this motivates the utilization of LLMs or speech large models for detecting and reconstructing corrupted speech. Nonetheless, a future work could refine conformal predictions by tackling calibration issues in lower-accuracy datasets (such as SVD and VOICED) or developing adaptive nonconformity scores to improve the precision of prediction sets.

Author Contributions: Conceptualization, P.G.-S., M.H. and V.V.-P.; Methodology, P.G.-S. and M.H.; Validation, M.H., P.G.-S. and V.V.-P.; Data curation, M.H.; Writing original draft preparation, M.H.; Writing review and editing, P.G.-S. and V.V.-P.; Supervision P.G.-S. All authors have read and agreed to the published version of the manuscript.

Funding: This research has been funded by the Spanish Ministry of Science, Innovation and Universities MICIU/AEI/10.13039/501100011033 under project number PID2023-147409NB-C21.

Data Availability Statement: The three datasets are available online and they are open-source. Kindly, you can download them from the following URLs (accessed on 5 May 2025):

- SLI: https://lindat.mff.cuni.cz/repository/xmlui/handle/11372/LRT-1597#
- SVD: https://github.com/HAMMADIPRO/SVD_downloader
- VOICED: https://physionet.org/content/voiced/1.0.0/

Conflicts of Interest: The authors declare no content conflicts of interest.

References

1. Sapienza, C.; Ruddy, B. *Voice Disorders*; Plural: San Diego, CA, USA, 2009.
2. Abdulmajeed, N.Q.; Al-Khateeb, B.; Mohammed, M.A. A review on voice pathology: Taxonomy, diagnosis, medical procedures and detection techniques, open challenges, limitations, and recommendations for future directions. *J. Intell. Syst.* **2022**, *31*, 855–875. [CrossRef]
3. Guyon, I.; Elisseeff, A. An introduction to variable and feature selection. *J. Mach. Learn. Res.* **2003**, *3*, 1157–1182.
4. Yang, X.S. Metaheuristic optimization: Algorithm analysis and open problems. In Proceedings of the International Symposium on Experimental Algorithms, Chania, Greece, 5–7 May 2011; Springer: Berlin/Heidelberg, Germany, 2011; pp. 21–32.
5. Pham, T.H.; Raahemi, B. Bio-inspired feature selection algorithms with their applications: A systematic literature review. *IEEE Access* **2023**, *11*, 43733–43758. [CrossRef]
6. Deb, K.; Pratap, A.; Agarwal, S.; Meyarivan, T. A fast and elitist multiobjective genetic algorithm: NSGA-II. *IEEE Trans. Evol. Comput.* **2002**, *6*, 182–197. [CrossRef]
7. Zitzler, E.; Laumanns, M.; Thiele, L. *SPEA2: Improving the Strength Pareto Evolutionary Algorithm*; ETH Zurich: Zürich, Switzerland, 2001.
8. Zhang, Q.; Li, H. MOEA/D: A multiobjective evolutionary algorithm based on decomposition. *IEEE Trans. Evol. Comput.* **2007**, *11*, 712–731. [CrossRef]
9. Grill, P.; Tučková, J. Speech databases of typical children and children with SLI. *PLoS ONE* **2016**, *11*, e0150365. [CrossRef]
10. Woldert-Jokisz, B. Saarbruecken Voice Database. 2007. Available online: https://stimmdb.coli.uni-saarland.de/ (accessed on 5 May 2022).
11. Cesari, U.; De Pietro, G.; Marciano, E.; Niri, C.; Sannino, G.; Verde, L. A new database of healthy and pathological voices. *Comput. Electr. Eng.* **2018**, *68*, 310–321. [CrossRef]
12. Tirumala, S.S.; Shahamiri, S.R.; Garhwal, A.S.; Wang, R. Speaker identification features extraction methods: A systematic review. *Expert Syst. Appl.* **2017**, *90*, 250–271. [CrossRef]

13. Angelopoulos, A.N.; Bates, S. Conformal prediction: A gentle introduction. *Found. Trends® Mach. Learn.* **2023**, *16*, 494–591. [CrossRef]
14. Shafer, G.; Vovk, V. A tutorial on conformal prediction. *J. Mach. Learn. Res.* **2008**, *9*, 371–421.
15. Junior, S.B.; Guido, R.C.; Aguiar, G.J.; Santana, E.J.; Junior, M.L.P.; Patil, H.A. Multiple voice disorders in the same individual: Investigating handcrafted features, multi-label classification algorithms, and base-learners. *Speech Commun.* **2023**, *152*, 102952.
16. Rehman, M.U.; Shafique, A.; Jamal, S.S.; Gheraibia, Y.; Usman, A.B. Voice disorder detection using machine learning algorithms: An application in speech and language pathology. *Eng. Appl. Artif. Intell.* **2024**, *133*, 108047. [CrossRef]
17. Zhang, M.; Wang, Y.; Zhang, W.; Yang, M.; Luo, Z.; Li, G. Inductive conformal prediction for silent speech recognition. *J. Neural Eng.* **2020**, *17*, 066019. [CrossRef] [PubMed]
18. Yadav, K. Automatic detection of vocal cord disorders using machine learning method for healthcare system. *Int. J. Syst. Assur. Eng. Manag.* **2024**, *15*, 429–438. [CrossRef]
19. Albadr, M.A.A.; Ayob, M.; Tiun, S.; AL-Dhief, F.T.; Al-Daweri, M.S.; Homod, R.Z.; Abbas, A.H. Fast learning network algorithm for voice pathology detection and classification. *Multimed. Tools Appl.* **2025**, *84*, 18567–18598. [CrossRef]
20. Sindhu, I.; Sainin, M.S. Automatic Speech and Voice Disorder Detection using Deep Learning—A Systematic Literature Review. *IEEE Access* **2024**, *12*, 49667–49681. [CrossRef]
21. Yildirim, M.; Kiziloluk, S.; Aslan, S.; Sert, E. A new hybrid approach based on AOA, CNN and feature fusion that can automatically diagnose Parkinson's disease from sound signals: PDD-AOA-CNN. *Signal Image Video Process.* **2024**, *18*, 1227–1240. [CrossRef]
22. Singh, N.; Sinha, S.; Singh, L. A novel WO-ANT: Whale-ant optimization algorithm for detection of Parkinson's disease. *Int. J. Inf. Technol.* **2024**, *17*, 1873–1881. [CrossRef]
23. Akila, B.; Nayahi, J.J.V. Parkinson classification neural network with mass algorithm for processing speech signals. *Neural Comput. Appl.* **2024**, *36*, 10165–10181. [CrossRef]
24. Hadjaidji, E.; Amara Korba, M.C.; Khelil, K. Improving Detection of Parkinson's Disease with Acoustic Feature Optimization Using Particle Swarm Optimization and Machine Learning. *Mach. Learn. Sci. Technol.* **2025**, *6*, 015026. [CrossRef]
25. Nasrolahzadeh, M.; Haddadnia, J.; Rahnamayan, S. Multi-objective optimization of wavelet-packet-based features in pathological diagnosis of alzheimer using spontaneous speech signals. *IEEE Access* **2020**, *8*, 112393–112406. [CrossRef]
26. Altay, E.V.; Alatas, B. Association analysis of Parkinson disease with vocal change characteristics using multi-objective meta-heuristic optimization. *Med. Hypotheses* **2020**, *141*, 109722. [CrossRef] [PubMed]
27. Sheth, P.; Patil, S. Evolutionary Jaya Algorithm for Parkinson's Disease Diagnosis using Multi-objective Feature Selection in Classification. In Proceedings of the 2019 5th International Conference On Computing, Communication, Control And Automation (ICCUBEA), Pune, India, 19–21 September 2019; IEEE: Piscataway, NJ, USA, 2019; pp. 1–4.
28. Patra, S.S.; Mittal, M.; Jena, O.P. Multiobjective evolutionary algorithm based on decomposition for feature selection in medical diagnosis. In *Predictive Modeling in Biomedical Data Mining and Analysis*; Elsevier: Amsterdam, The Netherlands, 2022; pp. 253–293.
29. McFee, B.; Raffel, C.; Liang, D.; Ellis, D.P.; McVicar, M.; Battenberg, E.; Nieto, O. librosa: Audio and music signal analysis in python. In Proceedings of the 14th Python in Science Conference, Austin, TX, USA, 6–12 July 2015; Volume 8, pp. 18–25.
30. Deb, K.; Kalyanmoy, D. *Multi-Objective Optimization Using Evolutionary Algorithms*; John Wiley & Sons, Inc.: New York, NY, USA, 2001.
31. Ishibuchi, H.; Masuda, H.; Tanigaki, Y.; Nojima, Y. Difficulties in specifying reference points to calculate the inverted generational distance for many-objective optimization problems. In Proceedings of the 2014 IEEE Symposium on Computational Intelligence in Multi-Criteria Decision-Making (MCDM), Orlando, FL, USA, 9–12 December 2014; IEEE: Piscataway, NJ, USA, 2014; pp. 170–177.
32. Biscani, F.; Izzo, D. A parallel global multiobjective framework for optimization: Pagmo. *J. Open Source Softw.* **2020**, *5*, 2338. [CrossRef]
33. Pedregosa, F.; Varoquaux, G.; Gramfort, A.; Michel, V.; Thirion, B.; Grisel, O.; Blondel, M.; Prettenhofer, P.; Weiss, R.; Dubourg, V.; et al. Scikit-learn: Machine learning in Python. *J. Mach. Learn. Res.* **2011**, *12*, 2825–2830.
34. Habib, M.; Vicente-Palacios, V.; García-Sánchez, P. Bio-inspired optimization of feature selection and SVM tuning for voice disorders detection. *Knowl.-Based Syst.* **2025**, *310*, 112950. [CrossRef]
35. Taquet, V.; Blot, V.; Morzadec, T.; Lacombe, L.; Brunel, N. MAPIE: An open-source library for distribution-free uncertainty quantification. *arXiv* **2022**, arXiv:2207.12274.

algorithms

MDPI

Article

Automatic Generation of Synthesisable Hardware Description Language Code of Multi-Sequence Detector Using Grammatical Evolution

Bilal Majeed [1,*,†], Rajkumar Sarma [1], Ayman Youssef [2], Douglas Mota Dias [3] and Conor Ryan [1,*]

[1] BDS Labs, Department of CSIS, University of Limerick, V94 T9PX Limerick, Ireland; rajkumar.sarma@ul.ie
[2] Department of Computers and Systems, Electronics Research Institute, Cairo 12622, Egypt; aymanmahgoub@eri.sci.eg
[3] Department of Computer Science & Applied Physics, Atlantic Technological University, H91 T8NW Galway, Ireland; douglas.motadias@atu.ie
[*] Correspondence: bilal.majeed@ul.ie (B.M.); conor.ryan@ul.ie (C.R.)
[†] Current address: T2-029, Lero, Tierney Building, University of Limerick, V94 NYD3 Limerick, Ireland.

Abstract: Quickly designing digital circuits that are both correct and efficient poses significant challenges. Electronics, especially those incorporating sequential logic circuits, are complex to design and test. While Electronic Design Automation (EDA) tools aid designers, they do not fully automate the creation of synthesisable circuits that can be directly translated into hardware. This paper introduces a system that employs Grammatical Evolution (GE) to automatically generate synthesisable Hardware Description Language (HDL) code for the Finite State Machine (FSM) of a Multi-Sequence Detector (MSD). This MSD differs significantly from prior work as it can detect multiple sequences in contrast to the single-sequence detectors discussed in existing literature. Sequence Detectors (SDs) are essential in circuits that detect sequences of specific events to produce timely alerts. The proposed MSD applies to a real-time vending machine scenario, enabling customer selections upon successful payment. However, this technique can evolve any MSD, such as a traffic light control system or a robot navigation system. We examine two parent selection techniques, Tournament Selection (TS) and Lexicase Selection (LS), demonstrating that LS performs better than TS, although both techniques successfully produce synthesisable hardware solutions. Both hand-crafted "Gold" and evolved circuits are synthesised using Generic Process Design Kit (GPDK) technologies at 45 nm, 90 nm, and 180 nm scales, demonstrating their efficacy.

Keywords: electronic design automation; evolvable hardware; sequence detectors; grammatical evolution; synthesisable sequential logic circuits; hardware description language design

1. Introduction

Designing digital circuits to meet specific requirements is a challenging task that demands significant effort from designers. This process is often time-consuming, and achieving an optimal design is not always guaranteed. To aid designers, EDA tools are used to enhance efficiency and produce more optimised circuits [1]. Some EDA tools incorporate Machine Learning (ML) techniques [2] or other forms of artificial intelligence [3,4] to assist in circuit design. However, current tools do not typically employ Evolutionary Algorithms (EA) to generate various circuit design solutions.

EAs are search-based techniques inspired by Darwinian evolutionary principles. They apply crossover and mutation, each with a set probability, to evolve toward an optimal solution for the given problem, mimicking natural selection to gradually enhance solution quality. The concept of genetic search was introduced by Alan Turing in 1948 [5]. Since then, the field of EAs has grown extensively, with applications in areas such as symbolic regression [6] and circuit design [7]. Prominent EA methodologies include Genetic Algorithms (GAs) [8], Genetic Programming (GP) [9], GE [10], and Evolutionary Strategies (ES) [11].

Evolvable Hardware (EH) refers to using EAs to evolve electronic circuits or their components. EH can be categorised into intrinsic and extrinsic evolution. Intrinsic evolution involves using hardware, such as Field Programmable Gate Arrays (FPGAs), to evolve and test solutions directly on the hardware in real time [12]. In contrast, extrinsic evolution relies on simulation tools to evolve and test circuits solely in a computer environment [13].

FPGAs are used to prototype digital circuits, which are specified at design time using HDLs, such as Very High-Speed Integrated Circuit HDL (VHSIC HDL or VHDL), Verilog, and SystemVerilog (SV), to describe circuit behaviour at different levels of abstraction. These languages enable designers to define circuits at both the gate level, using Boolean equations, and at the behavioural level, which is closer to high-level programming and includes constructs like loops and conditionals. The synthesis process then translates these descriptions into a gate-level netlist, which serves as the blueprint for physical implementation.

Digital sequential circuits, which differ from digital combinational circuits (detailed in Section 2) in their ability to store and manage state information, are critical in complex digital systems. These circuits are modelled using FSMs, which help represent the data flow and state transitions based on inputs. FSMs can be classified into two main types: Mealy machines and Moore machines (detailed in Section 2). In this work, we have evolved the MSD as a Mealy machine (shown in Figure 1) due to its ability to be highly responsive.

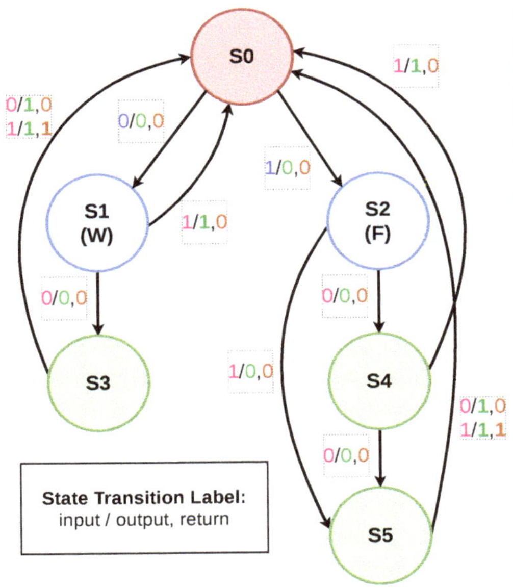

Type of Product/Payment	Associated Binary Value
Water (input)	'0'
Carbonated Drink (input)	'1'
50 ¢ (input)	'0'
1 € (input)	'1'
Payment Completion (output)	'0' / '1'
Return Change (output)	'0' / '1'

Figure 1. Mealy FSM of the single-input double-output Gold MSD.

To the authors' knowledge, no such MSD has been evolved in the literature using the evolutionary technique used in this work. As discussed in Section 4.1, GE is utilised to evolve this MSD and generate automatic HDL solutions. GE, a variant of GP that creates programs or expressions in virtually any syntax, uses context-free Backus–Naur Form (BNF) grammars for genotype-to-phenotype mapping. BNF is a formal notation used to describe the syntax of programming languages and data structures. It consists of production rules that define how non-terminal symbols can be expanded into sequences of terminal symbols and other non-terminals. More detail is given in Section 4.1.

These grammars allow the specification of rules that guide the evolution process. GE has been highly successful in various applications, including symbolic regression [14], classification [15], and particularly in automatic circuit design, whether combinational [16] or sequential [17], making it an ideal choice for this study. The decision to use behavioural-level HDL code instead of gate-level code for evolving this MSD is based on the recommendation for complex circuits [18] and its ability to evolve complex designs from the abstract level. Behavioural-level code is more straightforward to interpret [19] yet more challenging to evolve [20] due to its complex structures, such as *if–else* conditions and *for* loops.

MSDs are specialised sequential circuits that detect specific binary sequences, making them crucial in applications where sequence detection is critical, such as avionics and control systems. GE's use in evolving these circuits allows for the automatic generation of HDL solutions, particularly at the behavioural level. Due to its complex structures, HDL code is easier to interpret but more challenging to evolve.

This study demonstrates the practical application of this approach through a case study involving a vending machine, which requires effective state management for tasks like product selection and payment processing. While the specific application in this paper is that of vending machines, the technique can be applied to any MSD, such as those commonly found in traffic light control systems [21], elevator control systems [22], robot navigation systems [23], and TCP/IP communication protocols [24].

By evolving the HDL code for a Mealy FSM-based Gold MSD, which acts as a benchmark circuit, the research compares different parent selection techniques and verifies the synthesisability of the evolved designs. A Mealy machine is a type of FSM where the output depends on both the current state and the current input (detail in Section 2). This work contributes to the field by exploring well-established methods for automatic circuit design in the context of evolving complex sequential circuits, such as MSD, using behavioural-level HDL.

The main contributions of this work are summarised as follows:

- We propose a GE-based approach to evolve SV code for solution circuits representing the FSM of a single-input, multi-output MSD, mapped to a real-life vending machine.
- We provide a theoretical analysis showing that prior work focuses only on single-input, single-output, single-sequence detectors, and no existing work addresses the evolution of FSMs of this complexity.
- We conduct extensive experiments demonstrating that GE combined with LS and TS can effectively generate solution circuits for such real-world problems. However, LS performs better than TS in terms of time and computational resources.
- Although GE, LS, and TS are well-established techniques, the novelty of this work is reflected by putting them all together to evolve the solution circuits of real-life complex FSMs.
- We show that the evolved solution circuits are synthesisable, and their synthesis reports outperform those of manually designed circuits.

The organisation of this paper is as follows: Section 2 gives a detailed background on all the technical aspects used. Section 3 provides a literature survey on the evolution of digital circuits. Section 4 introduces GE and details the parent selection techniques employed in this study. Section 5 describes the process of generating training and test datasets necessary for the proposed work. Section 6 presents the experiments and results. Section 7 shows one of the perfectly evolved FSMs and compares it with a human-made FSM. Section 8 shows and compares the synthesis of the Gold and evolved solutions, highlighting the performance differences between human-designed and machine-generated solutions. Finally, Section 9 concludes the paper and discusses future directions in this field of research.

2. Background

HDLs are programming languages that describe digital circuits at the gate or behavioural (algorithmic) level. In gate-level descriptions, the circuit's architecture is defined using Boolean equations. In contrast, at the behavioural level, circuits are described using a high-level language similar to C, which includes constructs such as loops and if–else conditionals. Since we are working on the evolution of an FSM, no loops are used in this work, only conditionals, as the loops have no use here. Describing a circuit at the behavioural level significantly reduces the programmer's effort. The most widely used HDL standards are VHDL [25] and Verilog HDL [26]. SV [27], employed in this work to design and evolve sequential circuits, is an extension of Verilog HDL that enhances verification capabilities, earning it the designation of Hardware Verification Language (HVL) and an HDL.

Once a circuit's design is complete and its logical correctness verified through simulation, the next phase is logic synthesis. This process converts the HDL code into a netlist, which defines the circuit at the gate level using Boolean equations or registers. The netlist acts as a blueprint for constructing the circuit's logical structure. Tools like Cadence Genus [28] are used to optimise the synthesised design to meet specific criteria, including area, power consumption, system delay, and overall performance, based on predefined specifications such as operating temperature and transistor sizes targeting particular fabrication technologies.

However, even if a circuit passes simulation, physically constructing the chip might not be possible. This can occur if the netlist cannot be generated, often due to combining synthesisable and non-synthesisable constructs. For instance, loops in hardware must have known boundaries at compile time, even though they do not need them at simulation time. When loops have variable or non-static bounds, they cannot be synthesised. If we take the example of a `for` loop, the code will not be synthesised if the value of N is not known in `for (i = 0; i < N; i = i + 1)` at the synthesis time.

Digital circuits are broadly classified into two types: combinational circuits and sequential circuits. Combinational circuits produce outputs solely based on the current inputs, without memory of past inputs. In contrast, sequential circuits include memory elements that store information about the system's current state, making the output dependent on both the current state and the current input. Examples of basic sequential circuits include the JK Flip-Flop and the Up–Down Counter. Sequential circuits can be effectively modelled using FSMs. Because the output of a sequential circuit depends on its current state, FSMs are powerful tools for representing the data flow in terms of state transitions based on the inputs and the system's current state [29].

There are two primary types of FSMs: Mealy machines and Moore machines. In both types, the next state is determined by the system's current state and the circuit input. However, the critical difference lies in how the output is generated. In a Mealy machine, the output depends on the current state and input. In contrast, in a Moore machine, the output depends only on the current state. In this work, we are evolving the circuit based on Mealy FSM shown in Figure 1. We use Mealy machines because they respond quickly and typically require one state fewer than Moore machines, saving both time and hardware resources. However, Mealy machines can propagate transient input glitches—short, unintended signal fluctuations in a digital circuit—directly to the output if present. This is a trade-off we accept to achieve a high-speed and efficient system.

SDs vary in complexity depending on their size and features. As the name suggests, SDs detect specific binary sequences, such as '110011', and produce the desired output upon successful detection. They play crucial roles in real-time critical systems, such as avionics, where detecting a series of events in the correct order is essential. These systems must be highly responsive, meticulously designed, and thoroughly tested under all possible conditions to ensure they do not fail to generate necessary alarms.

This work proposes evolving behavioural-level HDL code using GE for a single-input, multi-output MSD. MSD is crucial in designing complex systems that require multiple states and transitions based on varying inputs. They are used in various applications where state management is critical, such as digital systems, embedded systems, and communication protocols. Speaking of their real-world applications, MSDs are fundamental in designing control systems for machinery, robotics, and other automated processes. Their ability to manage multiple states and transitions makes them suitable for real-time systems where efficient state management is crucial.

We apply this work to a real-life vending machine scenario, which provides a concrete example of how MSDs can be applied in real-world systems. Vending machines involve state management for product selection, payment processing, and item dispensing, making them an ideal test case for MSD designs. Moreover, the vending machine example highlights the complexity involved in real-world applications. It shows how MSDs can handle various states and transitions required to process inputs and provide outputs, demonstrating the practical value of evolving synthesisable HDL codes for such systems.

The Mealy FSM for the Gold MSD, representing a human-designed target circuit, is shown in Figure 1. "Gold" refers to the benchmark circuit the evolutionary system aims to achieve. This MSD can identify two distinct sequences, process a single 1-bit input, and produce two 1-bit outputs each clock cycle. The system starts at state S0 after a reset or upon waking from its idle state. The customer can choose between water and a carbonated drink. If the customer selects water (binary input '0'), the machine transitions to state S1 (labelled 'W'). Conversely, if the carbonated drink is selected (binary input '1'), the system moves to state S2 (labelled 'F'). After this selection, the system waits for payment. If the payment is incomplete within a specified time, the product is not dispensed, and the system resets to S0. We are assuming that the water costs EUR 1, while the carbonated drink costs EUR 1.50, with the system accepting only EUR 0.50 (binary input '0') and EUR 1 (binary input '1') coins.

In state S1, if the system receives EUR 1, it returns to S0, and the first output (leftmost) turns '1', indicating full payment and dispensing the water. However, if only EUR 0.50 is received, the system transitions to S3, keeping both outputs at '0' to indicate that additional payment is required. If another EUR 0.50 is received at S3, the system returns to S0, setting the first output to '1' to confirm full payment. If EUR 1 is received at S3, the system resets to S0 but also sets the rightmost output to '1', indicating that EUR 0.50 must be returned to the customer. For S2, the system transitions through states S4 and S5, eventually returning to S0 if three consecutive EUR 0.50 coins are inserted. If a EUR 1 coin is inserted at any point, the system bypasses certain states and returns to S0, providing appropriate outputs for payment processing and correct change. Note that in any given transaction, the maximum return possible is EUR 0.50.

As discussed above and shown in Figure 1, from any stage, there are two possible transitions: one dependent on input '1' and the other on input '0'. If, at any stage, the system does not receive any input—neither '1' nor '0'—it remains in the same state and outputs '0' for both product delivery and return change. An example of such a transition in the SV code of the human-designed FSM is shown in Figure 2 (full code given in Supplementary Material (SM), https://github.com/bmmajeed/MSD_VendingMachine (accessed on 2 June 2025), where the section's condition over the code responsible for this behaviour is highlighted in blue. These transitions are not explicitly shown in Figure 1 to prevent the diagram from becoming overly complex. However, in Section 7, it will be shown how these transitions affect the system's output, as the assignments within this `else` block are subject to evolution. Although the structure of the FSM is hardcoded, the system retains full flexibility to evolve all the assignments in the code.

```
always @ (posedge clk) // or posedge rst)
begin
        if (!rst)
        begin
                    ...
                    else if (state == S3)
                    begin
                        if (x1==0)
                        begin
                            state = S0;
                            out = 1;
                            rtrn = 0;
                        end
                        else if (x1==1)
                        begin
                            state = S0;
                            out = 1;
                            rtrn = 1;
                        end
                        else
                        begin
                            state = S3;
                            out = 0;
                            rtrn = 0;
                        end
                    end
                    ...
        end
        else
        begin
            state = S0;
            out = 0;
            rtrn = 0;
        end
end
```

Figure 2. Code snippet taken from human-made circuit of proposed FSM with the state's default assignments highlighted in blue colour.

3. Related Work

The origins of EH can be traced back to 1985 when Fourman et al. used GA to create compact symbolic circuit layouts [30], although the term "EH" was not used then. Subsequent studies, such as those by Cohoon et al. [31] and Kling et al. [32], employed GAs to optimally place components on circuit boards. In 1992, Koza [9] pioneered the use of GP to develop complete circuits at the gate level. This involved a decomposition method where an 11-to-1 multiplexer was evolved using two 6-to-1 multiplexers, which evolved from 3-to-1 multiplexers. The term "EH" was first introduced by [33] in 1993, marking a significant step towards the practical implementation of the "Darwin Machine" [34], where combinational circuits were evolved.

Considerable research has been conducted on the evolution of combinational circuits using various methods, including ES [35], Cartesian Genetic Programming (CGP) [36], and GE [16,37,38]. While significant work has been conducted on the evolution of FSMs or Deterministic Finite Automata (DFAs) [39–42], there has been less research on evolving sequential circuits, explicitly from the perspective of FSMs [43–46]. Notable examples include [47] and [17], where single-input, single-output SDs were evolved.

Ali et al. presented the first automatic solution for gate-level SD evolution using GA in 2004 [48]. They evolved a 4-bit SD ('1010') and a 6-bit SD ('011011') through a multi-stage evolutionary process, which was resource-intensive and did not directly produce the final solutions. The 6-bit SD combined two 3-bit SDs ('011'). The same circuits were re-evolved using GA, achieving more optimised solutions with reduced computational cost [49].

In 2007, Yao et al. evolved a 3-bit SD ('110') using an incremental, evolutionary approach based on GA [50]. This approach involved developing basic module circuits via a greedy search to construct more extensive, complex circuits. However, the approach needed to be expanded in scope, targeting specific modules and thus limiting the search space, preventing the generation of generalised solution circuits. In 2009, Xiong et al. evolved an overlapping 3-bit SD capable of detecting sequences like '101' and '100', utilizing intrinsic evolution on an FPGA board [51]. They later demonstrated this system's real-life application in a cardiovascular system [52], evolving the same SD using intrinsic and extrinsic evolution methods based on GA, showcasing the system on an FPGA board. Both approaches were computationally expensive for a small 3-bit SD.

In 2011, Soleimani et al. [47] revisited the evolution of the circuits initially evolved by Ali et al. [48] and Popa et al. [49] using a GA. They claimed to have used fewer gates than the earlier work by Ali et al., although they did not compare their results with Popa's work.

Tao et al. in 2012 evolved a 4-bit SD using GA and GP [53]. This incremental evolution process involved evolving basic circuits as modules before developing complex circuits from these modules. Using three evolutionary stages for a relatively simple 4-bit SD underscored the high computational cost.

Previous research has primarily focused on the gate-level evolution of SDs using GA and GP. The first work on behavioural-level SD evolution was presented in [17], where 3, 4, 5, and 6-bit SDs were evolved using GE. This work successfully evolved behavioural-level HDL codes from the perspective of FSMs, with the 3, 4, and 5-bit SDs being evolved in a single evolutionary stage. However, due to its complexity, the 6-bit SD required a second evolutionary phase incorporating grammar encapsulation. In this phase, the grammar is provided with the best individual achieved so far, allowing the system to evolve further from that point (detail in the [17]).

Building on this, we presented an improved performance in [44], where the same circuits were evolved using LS and a minimal set of test cases for training. This approach improved the performance of all circuits and enabled the evolution of the 6-bit SD in a single stage.

However, the studies above have focused exclusively on developing SDs for single-sequence detection; research has yet to explore the evolution of SDs capable of detecting multiple sequences. Additionally, while extensive research has been conducted on designing FSMs for vending machines [54–57], the evolutionary development of a vending machine-like FSM has not been investigated.

The evolutionary development of a vending machine-like FSM is a significant contribution and demonstrates the potential for automating the design of complex, real-world systems. This research demonstrates the capability of automating and optimising state management in digital systems by evolving FSMs for scenarios such as vending machines involving multiple states and decision points. This approach bridges theoretical methods with practical applications and paves the way for more efficient and innovative designs in various automated systems and consumer devices.

4. Evolutionary Computation Technique and Parent Selection Methods

4.1. Grammatical Evolution

GE [10] is a variation of GP that uses GAs to evolve variable-length binary strings, known as genotypes. These genotypes are mapped to programs or structures, referred to as phenotypes, in any language defined by a BNF grammar. This capability allows GE to evolve constructs in various programming languages.

BNF is a notation used to describe context-free grammars. Figure 3 illustrates the GE mapping process in detail. BNF grammars consist of four key components:

1. Production Rules (P): These define how a non-terminal symbol can be replaced with a combination of other symbols (terminals or non-terminals), guiding the generation of valid strings in the language defined by the grammar.
2. Terminals (T): These are the items that can appear in the final program and, in context-free grammars such as those used in this work, only on the Right-Hand Side (RHS) of a production rule and cannot be further expanded. Examples include symbols like '!' and '&' in the given grammar.
3. Non-Terminals (N): These are intermediate elements that are mapped by the production rules into members of the set T; these elements can appear on both the Left-Hand Side (LHS) and RHS of production rules and can be expanded into other non-terminals or terminals. For instance, $< \text{var} >$ is a non-terminal in the example.
4. Start Symbol (S): This is a special non-terminal from which the derivation begins. It appears in the first production rule and is typically the leftmost non-terminal, such as $< \text{expr} >$ in the first production rule shown in Figure 3.

The genotype, a binary string, is divided into *codons* and then converted into decimal integers. In the example shown in Figure 3, 8-bit codons of the genotype are used for this conversion. The first integer obtained is 40. The mapping process begins with the S from the first production rule (A). Given two possible RHS options in this rule, the selection is made using the modulus operation: the integer `40 modulo 2` equals 0, thus selecting the first option.

Next, the selected option contains a non-terminal, expanding the leftmost non-terminal. The second integer, 118, determines the selection within the non-terminal $< \text{var} >$, which has two options according to the third production rule (C). The operation `118 modulo 2` equals 0, so the first option, the variable 'x', is chosen. The subsequent non-terminal $< \text{op} >$ has three options according to the second production rule (B). Using the third integer, 124, the selection is made by calculating `124 modulo 3`, resulting in 1, thus choosing the second terminal, representing an OR operation. Finally, for the last non-terminal $< \text{var} >$, the integer `137 modulo 2` results in 1, selecting the variable 'y'. The resulting expression, shown in Figure 3, is an OR operation between the inputs 'x' and 'y'. This example is adapted from [17], where a more detailed explanation can be found.

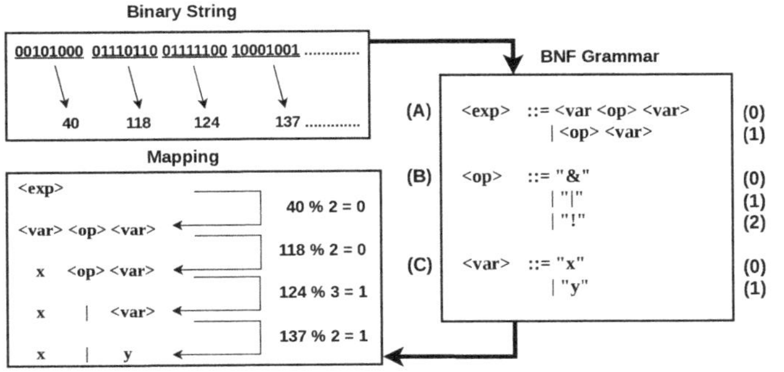

Figure 3. Genotype to phenotype mapping in GE.

4.2. Tournament vs. Lexicase Parent Selection

The parent selection method plays a critical role in the evolutionary process. In TS [58], a tournament is conducted among n individuals. The one with the best overall performance across all training cases, i.e., the highest aggregated fitness value, is selected. In this study, the value of n is set to 2 for all experiments to save computational time and cost. While this smaller tournament size increases the chances of low-fitness candidates being selected as parents and high-fitness candidates being missed [58], this effect is mitigated in our case due to the significant population sizes used. Larger populations increase diversity and the likelihood of high-fitness candidates being retained, ensuring robust evolutionary progress despite lower selection pressure.

In contrast, LS [59] does not rely on an aggregated fitness value. Instead, it selects specialists who excel at particular training cases. The process begins by considering the entire population, with individuals filtered out iteratively based on their performance on random training cases. The individual who performs best on the selected training case is chosen as a parent. One is randomly selected if two or more individuals perform equally well on a training case. LS has demonstrated exceptional problem-solving abilities in various fields, including regression [60] and evolutionary robotics [61]. Notably, it has shown outstanding results in the evolution of both combinational [20,37,38] and sequential [44] digital circuits in recent years.

5. Dataset Generation

The choice of training dataset is crucial in developing sequential circuits. The dataset can be random, exhaustive (where the system is initialised to a specific state and evaluated against all possible inputs), or selectively chosen based on the designer's preferences. The size or length of each training case within the dataset also plays a significant role.

In the FSM mapped to a vending machine, the number of training or test cases is limited, with each case having a specific length (the details are given below). The dataset shown in Figure 4, as shown below, includes all required training cases. In addition to having all the necessary training cases, it also represents the complete set of possible test cases for this FSM, as there are no additional scenarios applicable to this FSM. If the number of train and test cases is changed from those used in this work, it will increase the search space tremendously, and the trade-off would be a drastic decrease in success rate under a given setting of hyperparameters, which is not a useful trade-off. Since the training and test cases are the same, in Figure 4 they are indicated as TC. This contrasts with the more extensive test datasets used for testing other evolved SDs, such as those in [44,48]. If the train and test cases are kept different from each other and we have an unbalanced dataset, we can use a cluster-based approach, such as a hamming distance-based approach, to balance the dataset.

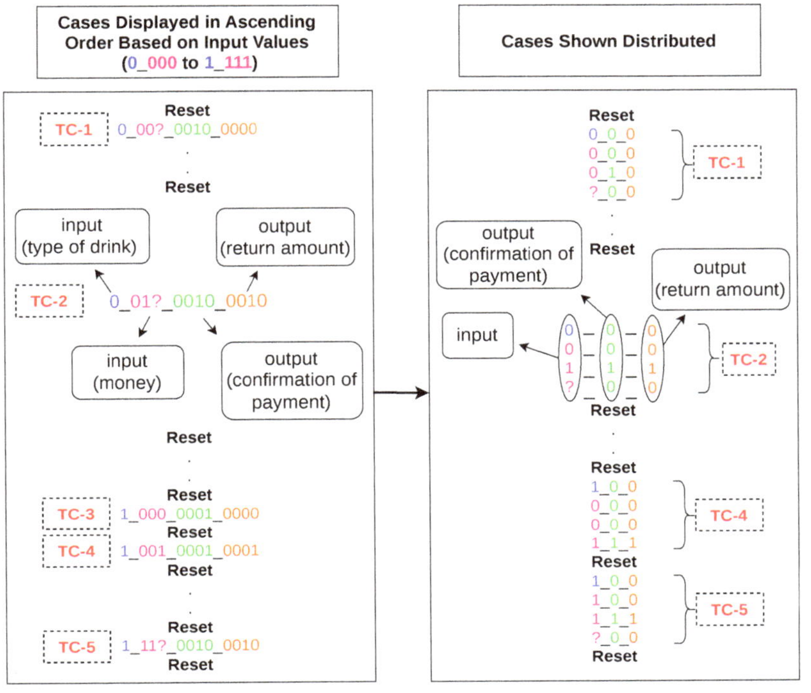

Figure 4. Training and test dataset design: The left side shows the 12-bit training or test case representation, which has inputs along with respective outputs in ascending order based on input values, while the right side shows the representation of each 12-bit case into four 3-bit vectors to show how they are fed into the system to compute the overall fitness using the SV testbench. TC stands for training or test case since the training cases are the only cases that can be used as test cases.

The TCs are presented in two formats for ease of explanation, as illustrated in Figure 4. On the left, the TCs are displayed input-wise, from '0_000' to '1_111', for clarity and ease of understanding. There are five example TCs shown in this figure, and they are selected as of mixed complexity to demonstrate the FSM's ability to handle both simple and extended input-output sequences involving varying payment paths and change-return scenarios. Take the example of the TC-2 '0_01?_0010_0010'. The higher order four bits (in this TC, '0_01?') represent inputs in product selection and payment status. The most significant bit indicates whether the customer selects water ('0') or a carbonated drink ('1'). The following three input bits immediately after the underscore ('_') represent what payment has been made for the selected product. As mentioned above, the water costs EUR 1, while a carbonated drink costs EUR 1.50. Therefore, these four bits indicate that the customer selected water, indicated by input '0', and then paid EUR 0.50 and EUR 1, indicated by '0' and '1', respectively. At this point, the payment is completed, and now the customer needs to be served with water and a return of EUR 0.50. In this scenario, the fourth bit in the input sequence of '0_01?' is taken as a don't care bit, shown as '?', and is neglected since the third input bit completes the payment. '?' indicates that the bit can be a '0', '1', or 'x' (no input at all) in this clock cycle. The same happens in TC-5, where the fourth bit in the input sequence is also neglected, but it does not always happen. It can be seen in TC-3 and TC-4 that after the fizzy drink ('1') is selected at the first bit of the input sequence, the payment is not completed by the end of the third bit in the input sequence, so the fourth bit is adding the value to the payment required to get the product delivered. Please refer to Figure 1 to understand this step in more detail.

The next two 4-bit chunks ('0010_0010' in TC-2), from left to right, in the training and test vectors, also separated by an underscore ('_'), represent the system's outputs: first confirming payment and the other returning any necessary change. The outputs are determined by combining the inputs and the system's state, as depicted in Figure 1 and discussed in Section 2. For input sequence TC-2 '0_01?', the first output sequence, '0010', with '1' at the third bit from left to right, reflects that the drink is served in the third cycle when the payment of EUR 1.50 is completed. This is because the consumer consumed the first two cycles, selecting the product and adding coins. Hence, the output is '0' in those two cycles. Since the payment of EUR 1.50 is more than the price of water, which is EUR 1, the '1' at the third bit from left to right in the following output sequence '0010' tells that in the third cycle, the customer has been paid with a change of EUR 0.50 along with the product.

The vending machine can only ever issue EUR 0.50 change. In this instance, EUR 0.50 was issued, and it happened in the same cycle as the drink being delivered. The change will always be issued in the same cycle as the delivered drink in the proposed system. Similarly, if there is no change, the final four bits will always be all zeroes. Since in the discussed TC here, the fourth bit in the input sequence is a don't-care bit, the outputs stay the same for the TCs with input sequences of '0_010', '0_011', or '0_01x'.

Note that we assume that the customer can only use EUR 0.50 and EUR 1 coins and that the vending machine will accept a specific amount of these coins in each cycle. Other coins are ignored.

On the right side of the dataset shown in Figure 4, each 12-bit TC from the left side is broken into four 3-bit vectors. Each vector has three bits, separated by '_', where the first bit indicates the input, and the remaining two bits represent the relevant outputs of the system for that input.

For example, the four input bits '0_011' from TC-2 on the left side are reflected in the leftmost bit of the four distributed vectors on the right side. Similarly, the four bits '0010' shown in the middle of TC-2 on the left side of Figure 4 are represented in the middle bit of the vectors on the right side of the figure. Finally, the four bits '0010' from the rightmost part of TC-2 on the left side are reflected in the rightmost bit of the distributed vectors on the right side. These three sections of the distributed vectors of TC-2 are enclosed in ovals and are properly labelled as shown on the right side of Figure 4 for a better understanding.

In this defined way, the TCs are distributed and fed into the system during each cycle, determining the system's fitness value against each TC. There are 16 12-bit TCs on the left side, which, when distributed according to the system's requirements, result in 64 3-bit vectors on the right side, which are fed into the system in each cycle. Thus, the maximum fitness value is 64. The maximum fitness value for an individual is only achieved if the individual correctly resolves all the TCs.

The pseudocode to generate the train/test dataset for a complex FSM, such as the one presented here, is shown in Algorithm 1. The process begins by initialising the FSM configuration, such as the vending machine logic discussed earlier. The input encoding is defined such that the first bit represents the product selection (with '0' for water and '1' for a fizzy drink), while the next three bits represent the payment sequence, where '0' and '1' correspond to EUR 0.50 and EUR 1 coins, respectively, and '?' denotes a don't-care condition. The output encoding is similarly defined, where the first four bits represent the delivery signal across clock cycles, and the next four bits represent the issuance of a return change if applicable. An empty dataset list is then initialised, and for each valid product and payment scenario, an input sequence is generated. The FSM transitions are simulated cycle-by-cycle to compute the corresponding delivery and return change outputs based on the system's logic and state evolution. Each TC is formatted by concatenating the input sequence, delivery output, and return change output into a 12-bit vector and is subsequently added to the dataset.

Algorithm 1 Generate Training Dataset for FSM Evolution.

```
function GENERATEDATASET
    Initialise FSM configuration                         // e.g., vending machine logic
    Define input encoding:
        1st bit: product selection                         // 0 = water, 1 = fizzy drink
        Next 3 bits: payment sequence       // 0 = €0.50, 1 = €1, ? = don't care
    Define output encoding:
        4 bits: delivery signal per cycle
        4 bits: return change signal per cycle
    Initialise empty dataset list D
    for all valid product and payment scenarios do
        Generate input sequence I                                          // 4 bits
        Simulate FSM state transitions for each cycle
        Compute outputs:
            Product delivery signal based on when payment completes
            Return change signal if overpayment occurs
        Format test case: TC = I · O_delivery · O_change
        Add TC to dataset D
    end for
    Initialise empty evaluation list V
    for all TC in D do
        Split TC (12 bits) into four 3-bit vectors:
        for i = 1 to 4 do
            v_i ← (input[i], delivery[i], change[i])
            Add v_i to V
        end for
    end for
    return V                          // 3-bit vectors for GE fitness evaluation
end function
```

In the distribution step tailored for the system's evaluation, each 12-bit TC is further split into four smaller vectors, each consisting of three bits. Each 3-bit vector comprises a single input bit and two associated output bits, corresponding to the delivery and return change signals for that particular cycle. These distributed vectors are then collectively used for fitness evaluation during the evolutionary process. The correct resolution of all vectors is essential for an individual to achieve the maximum fitness score. This structured approach ensures that all possible TCs relevant to the FSM are covered comprehensively, facilitating effective learning and circuit evolution within the GE framework.

While the dataset size is deliberately constrained to ensure complete coverage with minimal redundancy, the test complexity arises from the sequencing of state-specific inputs and strict timing constraints enforced by the FSM structure. Future work may explore extensional models, such as those proposed by Rodríguez et al. [62], to formally assess and adapt test complexity in broader, adaptive scenarios.

6. Experiments and Results

6.1. Experimental Setup, Tools, and Evolutionary Parameters

All experiments for this work were conducted using a setup that integrates Icarus Verilog (a Verilog/SV simulator) with Grammatical Algorithms in Python for Evolution (GRAPE) [63], a Python-based implementation of GE. The solutions evolved in GRAPE and were evaluated for each generation using Icarus Verilog to calculate the fitness value. The experiments were performed on a Dell OptiPlex 5070 (sourced from Dell Technologies Customer Solution Center, Limerick, Ireland) desktop computer with a 1 TB HDD, 256 GB SSD, a single 16 GB RAM module, and a 64-bit quad-core 9th-generation i7 processor with a 12 MB cache. The processor's standard operating frequency is 3.0 GHz, which can boost up to 4.7 GHz in turbo mode. The system uses the operating system Ubuntu 22.04.4 LTS.

The hyperparameters used in this work are detailed in Table 1. All experiments initially used a population of 1000 and ran for 30 generations. However, these parameters were increased for some experiments, with results presented according to the respective population size and maximum number of generations in Table 2. The experimental setup is designed to run 12 runs in parallel on 12 threads of the processor.

The grammar used in this study is shown in Figure 5. This work focuses on evolving the HDL code for a 6-state MSD, which is more complex than the previously evolved FSM for a 6-bit SD discussed in [44]. The increased complexity arises from the MSD's ability to track two sequences. We have evolved not only the system's next state and the values of two respective outputs—confirmation of payment and refund amount—but also the current state, input variables, and the values for input variables, which were not addressed in the previous work. This grammar facilitates the generation of the HDL module, which is simulated using Icarus Verilog.

The first production rule connects the *<parameters>* to *<sequential>*, where *<parameters>* defines hardcoded state parameters by assigning them 3-bit values. The *<sequential>* block contains the entire `always` block that implements the FSM. Within this `always` block, if the system is not undergoing a *reset*, the *<states_block>* selects the appropriate *if–else if* statement according to the system's current state (which is also evolved) and then enters *<conditional>*. The *<conditional>* block further selects the respective *if–else if* condition based on the evolved input variable and its evolved value, assigning the evolved next state and evolved outputs to the system. If the system is in a *reset* state, *<last_else>* is chosen within the `always` block, assigning hardcoded values to the next state and outputs, as specified in the grammar. The last three non-terminals only have terminals on the RHS, which are used throughout to evolve the current state, next state, input variable and its value, and the value of the outputs.

In this work, the primary evaluation metric used during the evolutionary process is the fitness score. The fitness score measures how accurately an evolved individual solves the target FSM problem by comparing its outputs against the expected outputs for a predefined set of TCs.

A total of 16 TCs divided into 64 3-bit vectors (as shown in Section 5) are used, where each case represents a specific input sequence and the corresponding expected output behaviour. For each 3-bit vector that the evolved individual resolves correctly, it is awarded a score of '1'. If the output for a vector does not match the expected behaviour, a score of '0' is assigned for that vector. The overall fitness score is computed by summing the individual scores across all 16 12-bit TCs or all 64 3-bit vectors. Therefore, the maximum achievable fitness score for an individual is '64', indicating perfect performance on all TCs.

```
<final>          ::= <parameters> <sequential>
<parameters>     ::= reg [2:0]state;
                     parameter S0 = 3'b000;
                                .
                                .
                     parameter S5 = 3'b100;
<sequential>     ::= always @(posedge clk)
                     begin
                         if(!rst) begin
                             <state_block>
                         end
                             <last_else>
                     end
<state_block>    ::= if (state == <state>)
                     begin
                         <conditional>
                     end
                     else if (state == <state>)
                     begin
                         <conditional>
                     end
                                .
                                .
                     <last_else>
<conditional>    ::= if (<var> == <val>)
                     begin
                         state = <state>;
                         out = <val>;
                         rtrn = <val>;
                     end
                     else if (<var> == <val>)
                     begin
                         state = <state>;
                         out = <val>;
                         rtrn = <val>;
                     end
                     else begin
                         state = <state>;
                         out = <val>;
                         rtrn = <val>;
                     end
<last_else>      ::= else begin
                         state = S0;
                         out = 0;
                         rtrn = 0;
                     end
<state>          ::= S0 | S1 | S2 | S3 | S4 | S5
<var>            ::= x0 | x1
<val>            ::= 0 | 1
```

Figure 5. BNF grammar with full structure of the HDL code for FSM to evolve the current state, input variables and their values, outputs' values, and the next state of the MSD.

The overall fitness score F for an individual is calculated by summing the scores across all 64 3-bit vectors and is defined as:

$$F = \sum_{i=1}^{N} s_i \quad (1)$$

where the following are used:

- $N = 64$ is the total number of 3-bit vectors;
- $s_i = 1$ if the individual's output for vector i matches the expected output;
- $s_i = 0$ otherwise.

Table 1. Hyperparameters used for experiments.

Hyperparameter	Value
No. of Runs	30
Population Size	1000
No. of Gens.	30
Initialisation	Sensible
Mutation Prob.	0.01
Crossover Prob.	0.9
Elitism	Yes
Training Cases	64
Test Cases	64

Table 2. Success rate and execution time of evolved SDs according to the LS and TS method with respective population size and number of generations. (NN stands for "Not Needed" and is used where a success rate of 25/30 or more is achieved and needs no more experiments. These success rates mark the successful runs and are highlighted in green. Unsuccessful runs are shown in blue).

Parameter Values		Training and Test Results Using LS		Training and Test Results Using TS	
Population Size	Number of Generations	Success Rate	Execution Time	Success Rate	Execution Time
1000	30	09/30	57 min 42 s	02/30	59 min 36 s
	100	17/30	178 min 41 s	06/30	177 min 42 s
	500	29/30	867 min 56 s	20/30	870 min 23 s
	1000	NN	NN	23/30	1732 min 53 s
5000	100	NN	NN	20/30	872 min 45 s
	500	NN	NN	30/30	4348 min 15 s

6.2. Experiments and Results Using Tournament Selection

In the first set of experiments, using TS as the parent selection method, the initial experiment was conducted with a population size of 1000 over 30 generations. The system took 59 min and 36 s to successfully evolve only two solutions across thirty runs, as reported in Table 2. Somewhat arbitrarily, we set a target of 25/30 successful runs; if a setup produces fewer than that, we consider the setup inadequate. In Table 2, blue text indicates the training and test success rate on unsuccessful runs, while green text indicates the training and test success rate greater than 25. No further experiments were run once we managed a success rate greater or equal to 25.

For this first experiment, the mean of the maximum fitness graph (all graphs for these experiments are provided in the SM, showing the mean maximum fitness scores of evolved circuits over 30 runs across the maximum number of generations) revealed consistently increasing error bars over the generations (as seen in the subsequent experiment shown in Figure 6a). Noting the significant variability in solutions indicated by the large error bars, the number of generations increased to 100, achieving a success rate of 6/30 and taking the execution time of 177 min and 42 s, almost three times the previous experiment. The resulting graph continued to show large error bars, as illustrated in Figure 6a. The number of generations was then further increased to 500, taking an execution time of 870 min and 23 s and resulting in a success rate of 20/30. Although the results did not improve fivefold with the increased number of generations, the error bars remained substantial.

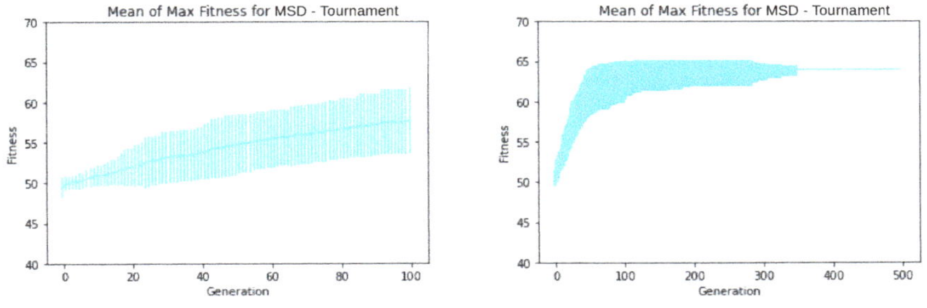

(**a**) Pop. = 1000, Gens. = 100 (**b**) Pop. = 5000, Gens. = 500

Figure 6. Error bar graphs representing the mean of the maximum fitness values of the individuals across experiments run with the mentioned population size and number of generations using TS.

The number of generations was further increased to 1000, resulting in a success rate of 23/30, taking execution time of 1732 min and 53 s. The error bars on the graph were significantly reduced. We experimented with swapping these values to explore whether a larger population size with fewer generations might yield better results. The population size was set to 5000, and the number of generations was reduced to 100 (thus keeping the total number of individuals the same). With these settings, we achieved the same success rate of 20/30, with an execution time of 872 min and 45 s. However, the error bars indicated considerable variability in the solutions, with no clear trend toward saturation.

We then kept the population size at 5000 and increased the number of generations to 500. This configuration took execution time of 4348 min and 15 s and led to a perfect success rate of 30/30, with the system reaching this rate around 350 generations, as shown in Figure 6b.

Since the dataset used for training is the only one available for the test success rate, the entire dataset was used for training and testing. The respective success rate of the evolved solutions on the test dataset is indicated in the same column in Table 2, showing perfect accuracy for all evolved solutions.

6.3. Experiments and Results Using Lexicase Selection

All settings were kept consistent with the TS method, and the first experiment was conducted with a population size of 1000 across 30 generations, repeated over 30 runs. This resulted in a training and test success rate of 09/30, taking an execution time of 57 min 42 s, as shown in Table 2. Since this is not an acceptable success rate, we increased the number of generations to 100. This adjustment led to a success rate of 17/30, with an execution time of 178 min and 41 s, which is approximately three times higher than the TS method under the same generation setting, as seen in Table 2. Despite this improvement, the graph still displayed large error bars and a lack of saturation, leading to a further increase in the number of generations to 500. This change resulted in a success rate of 29/30, taking an execution time of 867 min and 56 s, again significantly outperforming the TS method under identical hyperparameter settings. No further experiments were conducted because this success rate was nearly perfect. The resulting graph showed the mean of the maximum fitness value approaching saturation, with consistently narrowing error bars.

It can be seen that in evolving FSMs, where multiple diverse and sometimes conflicting TCs must be satisfied, LS outperforms TS by preserving individuals that solve specific difficult cases. Unlike TS, which favours generalist individuals with good average performance and risks losing useful specialists early, LS evaluates performance case-by-case, maintaining a richer diversity of FSM behaviours. This diversity is crucial for complex FSM evolution, where different transitions and outputs may be optimised at different stages. As a result, LS enables better exploration and often leads to higher-quality FSM solutions.

Incorporating both LS and TS, the proposed method is highly scalable to larger and more complex FSMs. However, scaling may require greater computational and evolutionary resources. To adapt this methodology for larger FSMs, modifications to the grammar, training dataset, and testbench would be necessary. For example, expanding the current FSM to support more than two products would require a complete restructuring of the dataset and FSM, alongside corresponding updates to the grammar and testbench. Future work will focus on scaling this methodology to enable the evolution of more complex and larger FSMs.

7. Evolved Solution FSM

7.1. Code and Transition Analysis

FSM of one of the perfectly evolved solution circuits in SV is shown in Figure 7. This FSM is taken from the solution circuits generated using LS, and the experiment has a top success rate of 29/30. In this evolved FSM, all state transitions and their conditions are automatically generated through the evolutionary process discussed earlier. The FSM structure includes six states, named S0 to S5, aligning with the human-designed FSM presented in Section 2. Notably, the evolved solution exhibits a broader set of transitions, showcasing the creative and exploratory nature of the GE.

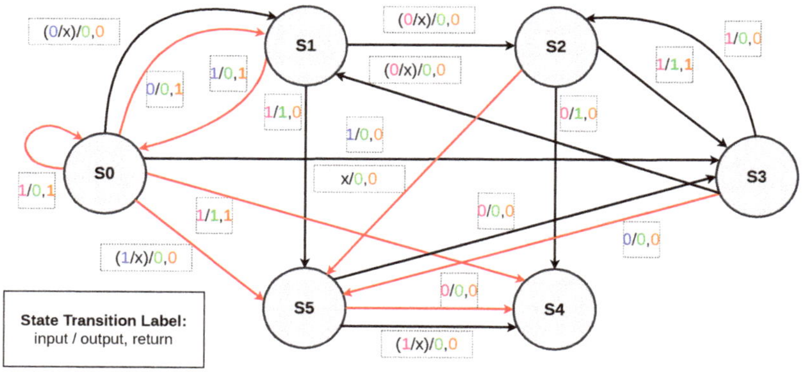

Figure 7. FSM of one of the perfectly evolved solution circuits. (Black arrows showing the active transitions, while the red arrows show extraneous transitions).

The FSM shows transitions in both black and red colours. The black transitions are the ones in use that occur during execution based on the testbench inputs. The red transitions exist in the code but are never triggered during real operation. These extra transitions show how the evolved system can explore a wider range of possible behaviours and add flexibility to the design.

In this structure, transitions originating from state S0 based on x1, indicated in dark pink within the transition labels, are not functionally utilised, as x1 is not active during S0 according to the testbench configuration. These transitions are marked in red in the FSM diagram to indicate that they remain inactive in this context. Similarly, transitions from states S1 through S5 based on x0, represented in blue within the transition labels, are also non-operative in actual execution, as x0 is only defined during the initial cycle in S0. The colour scheme used for input bits in the transition labels is consistent with the human-designed FSM in Figure 1, aiding in visual continuity and interpretability.

Some transitions are labelled with combined conditions such as '(0/x)' or '(1/x)'. These originate from broader `else` blocks in the evolved code, as opposed to the more precise `else if` chains used in the human-designed solution. While this approach may appear less constrained, it highlights the flexibility of the evolved FSM to accommodate various input combinations without compromising functionality.

An interesting aspect of the evolved FSM is the presence of repeated or overlapping state conditions. For instance, as shown in Figure 8, state S0 is defined in two separate blocks, both beginning with `state == S0`. Given the sequential nature of the code, only the first such block is ever executed during runtime, while the second block (highlighted in red) remains extraneous. This is reflected in the FSM diagram in Figure 7, where transitions linked to the second block are highlighted in red. These extra structures, although inactive, are byproducts of the evolutionary search process and illustrate the system's openness to exploring alternative representations.

Additionally, certain conditional checks, such as `x1 == 1`, appear in S0, despite x1 being undefined during that state. It can be seen in Figure 8 that the conditional statement based on `x1 == 1` in the first block based on S0 will never execute, yet is highlighted in red. The remaining part of this block, which will be able to execute, is highlighted in green. Likewise, some input checks like `x1 == 0` occur multiple times within the same state, as can be seen in the transitions from S5, where two transitions are based on the same input and output labels. However, one is valid and the other is not. These repetitions and overlaps are not errors but rather emergent outcomes of the grammar-based evolutionary system, which emphasises functionality and completeness over stylistic minimalism. They underscore the potential of GE to generate circuits that are not only correct but also adaptable and structurally diverse.

Thus, while the evolved FSM performs correctly on all evaluated TCs, its code also reflects the flexibility and exploratory nature of the evolutionary process, resulting in additional transitions and structures that, although not activated under current testbench constraints, demonstrate the system's capacity to adapt and generalise beyond strictly defined scenarios.

7.2. Test Case Mapping onto the Evolved FSM

After the FSM is evolved, its functionality is validated by mapping the six TCs (five of which were defined earlier in Section 5 and shown in Figure 4, the dataset generation process), onto the evolved state machine. These TCs represent various user interactions with the vending machine, covering all valid input-output behaviours. Each TC is shown in a unique colour on the FSM diagram in Figure 9 to clearly highlight the path it follows through the states.

```
always @ (posedge clk) // or posedge rst)
begin
        if (!rst)
        begin
                ...
                else if (state == S0) begin
                    if (x1 == 1) begin
                        state = S4;
                        out = 1;
                        rtrn = 1;
                    end
                    else if (x0 == 1) begin
                        state = S3;
                        out = 0;
                        rtrn = 0;
                    end
                    else begin
                        state = S1;
                        out = 0;
                        rtrn = 0;
                    end
                end
                else if (state == S0) begin
                    if (x1 == 1) begin
                        state = S0;
                        out = 0;
                        rtrn = 1;
                    end
                    else if (x0 == 0) begin
                        state = S1;
                        out = 0;
                        rtrn = 1;
                    end
                    else begin
                        state = S5;
                        out = 0;
                        rtrn = 1;
                    end
                end
                ...
        end
        else
        begin
            state = S0;
            out = 0;
            rtrn = 0;
        end
end
```

Figure 8. Code snippet taken from one of the perfectly evolved solution circuits using LS.

The mapping uses the following colour scheme for the six TCs:

- Azure for TC-1;
- Orange for TC-2;

- Maroon for TC-3;
- Purple for TC-4;
- Citrine yellow for TC-5;
- Parrot green for TC-6 (an additional test case not shown in the original dataset figure).

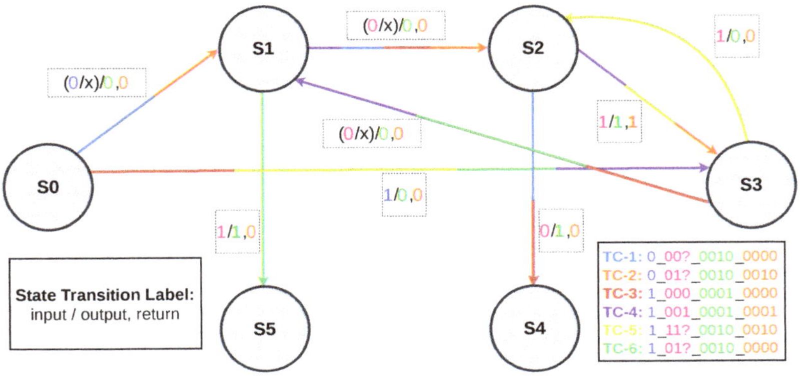

Figure 9. Mapping of TCs onto evolved FSM.

Each coloured path corresponds to a valid sequence of transitions based on the inputs defined in the respective test case. For example, in TC-2, the system begins in state S0, detects a water selection x0 = 0, and proceeds to state S1. From there, it transitions through additional states depending on the payments made using x1, until the total payment matches the product cost.

The same logic applies to the fizzy drink selection path, such as in TC-4, which begins with x0 = 1 in S0 and continues through states like S3, S1, and S2, depending on the combination and order of EUR 0.50 and EUR 1 coins. It finally ends at S3 with the delivery of the product and a return change of EUR 0.50. In each TC, the state transitions and output responses follow the correct timing and behaviour, as expected from the FSM.

The additional TC, TC-6, is shown in parrot green in Figure 9 and follows the input and output pattern '1_01x_0010_0000'. This TC is specifically included to highlight the use of state S5, which is not involved in the first five TCs. By showing a path that passes through S5, the evolved FSM demonstrates its coverage of all relevant states and confirms its flexibility in handling less common but still valid scenarios. The transition labels in this figure also follow the same colour coding as used in the dataset generation figure (Figure 4).

Even though the evolved FSM shown in Figure 7 includes some transitions that are not used during execution, these do not affect the system's operation. The testbench restricts the inputs per cycle, so only the correct transitions are activated. The evolved FSM, therefore, completes all TCs correctly and achieves the maximum fitness score of 64, meaning it delivers the right outputs for all 64 input vectors derived from the dataset.

8. Logic Synthesis

After successfully generating HDL codes for MSDs using both TS and LS, the evolved solutions (successful candidates from each run) were synthesised and compared with the synthesis of the Gold Circuit. For this, 29 and 30 solution circuits, which are the best of the run from each of LS and TS, respectively, where the best success rates were achieved, were taken towards the synthesis, as shown in Table 2. The synthesis process employed Complementary Metal–Oxide–Semiconductor (CMOS) GPDK technologies at 45 nm, 90 nm, and 180 nm nodes, using the Cadence Genus synthesis tool at a clock frequency of 100 MHz. The reset signal was configured to the ideal network in the constraints file. For all three GPDK technologies, synthesis was performed under both Fast Corner (FC) and Slow Corner (SC) conditions. FC represents the most optimal operational environment for the circuit, based on specific temperature and supply voltage settings, while SC denotes the least favourable conditions. The temperature and voltage values used in these designs are detailed in Table 3.

Table 3. Nominal temperature and supply voltage for logic synthesis of relevant technologies.

Parameters	45 nm		90 nm		180 nm	
	FC	SC	FC	SC	FC	SC
Temperature (°C)	0	125	0	125	0	125
Voltage (V)	1.32	1.08	1.1	0.9	1.98	1.62

Table 4 presents a comparison of synthesis reports between the Gold MSD and the MSDs evolved using TS and LS, focusing on area (measured in the number of cells) and Power Delay Product (PDP). The PDP, calculated as the product of power and delay, is a crucial metric in evaluating IC design performance, given the inherent trade-offs between power consumption and delay. The unit of PDP is milliwatt-picoseconds (mWps). The comparisons in Table 4 are organised by the relevant GPDK technology, highlighted in three different shades of maroon; the lighter the shade, the smaller the GPDK technology. For instance, the synthesis reports of the Gold Circuit under FC and SC conditions using 45 nm technology are compared against those of the evolved circuits using the same technology under TS and LS. This method is similarly applied to 90 nm and 180 nm technologies.

For each technology, the Gold Circuit's metrics under FC and SC conditions are compared with the best (minimum), mean/average (AVG), median, and worst (maximum) values among the 29 perfect solution circuits evolved through LS and the 30 solution circuits evolved through TS. The table also includes the standard deviation (STD) among all values. At least one synthesised solution circuit for each technology generated through the proposed system has performed either better than the Gold Circuit, highlighted in green, or equal to the Gold Circuit, highlighted in blue. It can be seen that only one value out of 24 values is highlighted in blue, and the rest are shown in green, which means that evolution has produced a better result than the Gold Circuit under almost every circumstance.

Moreover, to show if there is a statistically significant difference between the synthesis reports of gold and evolved circuits, p-values extracted from the statistical hypothesis test named one-sample t-test are also given, where the Gold Circuits are compared with evolved circuits. Since the p-value if less than or equal to 0.05 indicates that there is a significant difference between the Gold Circuit and the group of circuits evolved under the specific technology, it is shown in bold and orange colour that for 20 out of 24 values of area and PDP values taken from evolved circuits, the null hypothesis can be rejected. There is a significant difference compared with the values of the Gold Circuit. There are only four values where the null hypothesis cannot be dismissed, and we have to note that there is no significant difference between the Gold Circuit's values and the mean values taken from the group of 30 or 29 circuits for a specific technology. Given this, we can safely say that most of the circuits are significantly different and better than the Gold Circuit, and we can generate better solutions automatically through evolutionary ML, which is the primary goal of this work.

Solutions evolved through LS achieve a higher success rate compared with TS. The performance of the synthesised designs from LS also surpasses those from TS, since three out of four values with a p-value greater than 0.05 belong to TS. This leads the authors to recommend LS as the preferable parent selection technique for evolving complex sequential circuits.

Table 4. A comparison of synthesis reports generated using GPDK 45 nm, 90 nm, and 180 nm on FC and SC in terms of the best (minimum), average, standard deviation, median, worst (maximum), and statistical hypothesis test's p-values of area (measured in the unit of number of cells) and PDP (measured in the unit of milliwatt-picoseconds) between the Gold and evolved circuits using TS and LS.

Gold Circuit	Parameter	45 nm		90 nm		180 nm	
		FC	SC	FC	SC	FC	SC
	Area	19	26	20	20	17	17
	PDP	115.7	80.56	64.51	79.2	356.5	483.1

	Parameter	Evolved Through TS											
		FC						SC					
		Best	Mean	STD	Med.	Worst	p-value	Best	Mean	STD	Med.	Worst	p-value
45 nm	Area	18	24.66	4.84	25	37	5.26e-07	21	29.3	4.78	30	40	7.2e-04
	PDP	106.1	126.9	11.08	125.85	157.46	5.44e-06	72.08	88.07	9.26	87.64	110.6	1.21e-04
	Parameter	Evolved Through LS											
		FC						SC					
		Best	Mean	STD	Med.	Worst	p-value	Best	Mean	STD	Med.	Worst	p-value
	Area	12	25.93	4.52	26	33	5.47e-09	19	31.03	4.52	31	38	1.86e-06
	PDP	110.8	135.62	14.81	131.96	160.78	6.92e-08	66.6	94.95	13.47	93.6	117.15	3.57e-06

	Parameter	Evolved Through TS											
		FC						SC					
		Best	Mean	STD	Med.	Worst	p-value	Best	Mean	STD	Med.	Worst	p-value
90 nm	Area	17	24.4	4.03	24.5	33	1.69e-06	17	24.4	4.08	24.5	33	2.07e-06
	PDP	58.3	64.47	4.14	63.38	77.67	0.96	72.21	79.75	5.79	78.32	96.81	0.607
	Parameter	Evolved Through LS											
		FC						SC					
		Best	Mean	STD	Med.	Worst	p-value	Best	Mean	STD	Med.	Worst	p-value
	Area	13	25.55	4.29	25	33	1.46e-07	13	25.52	4.19	25	33	1.02e-07
	PDP	57.2	67.4	7.40	64.75	81.48	0.047	70.54	83.42	9.56	80.04	101.58	0.025

	Parameter	Evolved Through TS											
		FC						SC					
		Best	Mean	STD	Med.	Worst	p-value	Best	Mean	STD	Med.	Worst	p-value
180 nm	Area	17	23.5	3.86	23.5	31	4.14e-10	16	23.7	3.92	24	31	2.93e-10
	PDP	317.9	365.6	24.49	360.77	440.1	0.05	404.8	471.6	40.57	471.9	596.42	0.13
	Parameter	Evolved Through LS											
		FC						SC					
		Best	Mean	STD	Med.	Worst	p-value	Best	Mean	STD	Med.	Worst	p-value
	Area	12	24.45	4.28	24	33	3.98e-10	16	24.1	3.78	24	32	7.52e-11
	PDP	319.6	377.21	41.84	369.2	454.78	0.013	418.5	503.1	59.34	483.47	614.37	0.08

A two-sample t-test is run between the synthesis reports of the groups of circuits evolved through LS and TS, and the extracted p-values are shown in Table 5. The values less than 0.05 show a significant difference between the two groups, shown in bold and red, while the rest are in bold and green. So, it can be seen that although LS performed much better than TS in terms of evolutionary success rate, the circuits evolved through both are pretty much similar according to the synthesis reports, yet acceptable.

Table 5. T-test results run on the groups of solution circuits evolved through LS and TS under the relevant GPDK technologies.

Parameter	45 nm				90 nm				180 nm			
	FC		SC		FC		SC		FC		SC	
	Area	PDP	Area	PDP	Area	PDP	Area	PDP	Area	PDP	Area	PDP
p-value	0.304	0.014	0.157	0.027	0.293	0.077	0.304	0.082	0.376	0.202	0.689	0.022

9. Conclusions and Future Work

This work introduces the evolution of synthesisable HDL codes for a single-input, multi-output MSD using GE and two parent selection methods: TS and LS. Although these parent selection methods and GE are well-established techniques, they are combined here to evolve the FSM of a real-life MSD, which is complex and hard to evolve and speaks to the novelty of the presented system. The evolved MSD is applied to a practical vending machine scenario, covering the entire process from product selection through payment to delivery. Using the evolved MSD designs in a vending machine scenario, the study validates the effectiveness of the evolutionary methods in generating practical, real-world solutions. It also shows the potential for these methods in other complex systems requiring similar state management capabilities, such as robotics.

The results demonstrate that LS significantly outperforms TS in evolving a sequential Mealy machine in terms of success rate and computational efficiency. The Gold standard and evolved circuits are synthesised to validate the practical application using GPDK at 45 nm, 90 nm, and 180 nm. These syntheses confirm that the circuits evolved to represent a significant step forward in automated chip design and manufacturing. Notably, at least one evolved circuit outperforms the Gold Circuit using either TS or LS. This study is the first to evolve synthesisable HDL codes for MSDs.

Future work will focus on evolving more complex circuits, such as multi-input and multi-output MSDs. In addition, another future direction will consider exploring the systematic hyperparameter tuning techniques to optimise evolutionary ML approaches, such as GE, which will likely improve overall system performance. We will also investigate using some other implementations of GE than GRAPE, such as Grammatical Algorithms in C++ for Evolution (GRACE), where the LS can be compared with the results of the approach used here.

Supplementary Materials: The following supporting information can be downloaded at: https://www.mdpi.com/article/10.3390/a18060345/s1, Data_Set_Paper: Figure showing dataset generation used for training the model; Lexicase_Selection_Experiment_1: Mean of average fitness across generations for first experiment using Lexicase parent selection; Lexicase_Selection_Experiment_2: Mean of average fitness across generations for second experiment using Lexicase parent selection; Lexicase_Selection_Experiment_3: Mean of average fitness across generations for third experiment using Lexicase parent selection; Tournament_Selection_Experiment_1: Mean of average fitness across generations for first experiment using Tournament parent selection; Tournament_Selection_Experiment_2: Mean of average fitness across generations for second experiment using Tournament parent selection; Tournament_Selection_Experiment_3: Mean of average fitness across generations for third experiment using Tournament parent selection; Tournament_Selection_Experiment_4: Mean of average fitness across generations for fourth experiment using Tournament parent selection; Tournament_Selection_Experiment_5: Mean of average fitness across generations for fifth experiment using Tournament parent selection; Tournament_Selection_Experiment_6: Mean of average fitness across generations for sixth experiment using Tournament parent selection; Grammar.txt: BNF Grammar used for evolution; Human_Designed_Module.sv: SV code of human-designed module in SV file format; Human_Designed_Module.txt: SV code of human-designed module in text file format; Evolved_Solution.sv: SV code of machine-designed module in SV file format; Evolved_Solution.txt:

SV code of machine-designed module in text file format; Dataset.txt: 12-bit 16 train and test vectors used for training and test purposes.

Author Contributions: Conceptualisation, B.M., R.S., A.Y., D.M.D. and C.R.; methodology, B.M., A.Y., R.S. and C.R.; formal analysis, B.M., R.S. and C.R.; data curation, B.M.; writing—original draft preparation, B.M.; writing—review and editing, R.S., A.Y., D.M.D. and C.R.; visualization, B.M., R.S. and C.R.; supervision, R.S., A.Y., D.M.D. and C.R.; project administration, D.M.D., R.S. and C.R.; funding acquisition, C.R. All authors have read and agreed to the published version of the manuscript.

Funding: This work was supported by the Research Ireland (Science Foundation Ireland (SFI) before) grant 16/IA/4605.

Data Availability Statement: The dataset used in this work can be accessed at Supplementary Materials.

Conflicts of Interest: The authors declare no conflicts of interest. The funders had no role in the design of the study; in the collection, analyses, or interpretation of data; in the writing of the manuscript; or in the decision to publish the results.

Abbreviations

The following abbreviations are used in this manuscript:

GE	Grammatical Evolution
HDL	Hardware Description Language
FSM	Finite State Machine
MSD	Multi-Sequence Detector
SD	Sequence Detector
TS	Tournament Selection
LS	Lexicase Selection
GPDK	Generic Process Design Kit
EDA	Electronic Design Automation
ML	Machine Learning
EA	Evolutionary Algorithm
GA	Genetic Algorithm
GP	Genetic Programming
ES	Evolutionary Strategies
EH	Evolvable Hardware
FPGA	Field Programmable Gate Arrays
VHDL	Very High-Speed Integrated Circuit HDL
SV	SystemVerilog
HVL	Hardware Verification Language
BNF	Backus–Naur Form
CGP	Cartesian Genetic Programming
DFA	Deterministic Finite Automata
P	Production Rules
T	Terminals
N	Non-Terminals
S	Start Symbol
RHS	Right-Hand Side
LHS	Left-Hand Side
GRAPE	Grammatical Algorithms in Python for Evolution
NN	Not Needed
SM	Supplementary Material
CMOS	Complementary Metal–Oxide–Semiconductor
FC	Fast Corner
SC	Slow Corner
PDP	Power Delay Product
mWps	milliwatt-picosecond
AVG	Average
STD	Standard Deviation

References

1. Farrahi, A.; Hathaway, D.; Wang, M.; Sarrafzadeh, M. Quality of EDA CAD tools: Definitions, metrics and directions. In Proceedings of the IEEE 2000 First International Symposium on Quality Electronic Design (Cat. No. PR00525), San Jose, CA, USA, 20–22 March 2000; pp. 395–405.
2. Solido. Solido Design Solutions. 2005. Available online: https://eda.sw.siemens.com/en-US/ic/solido/ (accessed on 1 November 2022).
3. Eagle. Eagle by Autodesk. 1988. Available online: https://www.autodesk.com/products/eagle/overview (accessed on 1 November 2022).
4. Kicad. KiCad Electronic Design Automation. 1992. Available online: https://www.kicad.org/ (accessed on 1 November 2022).
5. Ince, D.C. *Mechanical Intelligence (Collected Works of A. M. Turing)*; North-Holland Publishing Co.: Amsterdam, The Netherlands, 1992.
6. Zelinka, I.; Oplatkova, Z.; Nolle, L. Analytic programming–Symbolic regression by means of arbitrary evolutionary algorithms. *Int. J. Simul. Syst. Sci. Technol.* **2005**, *6*, 44–56.
7. Nicosia, G.; Rinaudo, S.; Sciacca, E. An evolutionary algorithm-based approach to robust analog circuit design using constrained multi-objective optimization. In *Research and Development in Intelligent Systems XXIV*; Springer: London, UK, 2007; pp. 7–20.
8. Mirjalili, S. Genetic Algorithm. In *Evolutionary Algorithms and Neural Networks: Theory and Applications*; Springer International Publishing: Cham, Switzerland, 2019; pp. 43–55.
9. Koza, J.R. *Genetic Programming: On the Programming of Computers by Means of Natural Selection*; MIT Press: Cambridge, MA, USA, 1992.
10. Ryan, C.; Collins, J.J.; Neill, M.O. Grammatical evolution: Evolving programs for an arbitrary language. In Proceedings of the European Conference on Genetic Programming, Paris, France, 14–15 April 1998; Springer: Berlin/Heidelberg, Germany, 1998; pp. 83–96.
11. Rudolph, G. Evolutionary Strategies. In *Handbook of Natural Computing*; Springer: Berlin/Heidelberg, Germany, 2012; pp. 673–698.
12. Zhang, Y.; Smith, S.; Tyrrell, A. Digital circuit design using intrinsic evolvable hardware. In Proceedings of the 2004 NASA/DoD Conference on Evolvable Hardware, Seattle, WA, USA, 26 June 2004; pp. 55–62.
13. Kalganova, T. An Extrinsic Function-Level Evolvable Hardware Approach. In Proceedings of the Genetic Programming, Edinburgh, UK, 15–16 April 2000; Poli, R., Banzhaf, W., Langdon, W.B., Miller, J., Nordin, P., Fogarty, T.C., Eds.; Springer: Berlin/Heidelberg, Germany, 2000; pp. 60–75.
14. Ali, M.; Kshirsagar, M.; Naredo, E.; Ryan, C. Towards Automatic Grammatical Evolution for Real-world Symbolic Regression. In Proceedings of the 13th International Joint Conference on Computational Intelligence, Virtual, 25–27 October 2021; Volume 1: ECTA, INSTICC, pp. 68–78.
15. Murphy, A.; Murphy, G.; Amaral, J.; Mota Dias, D.; Naredo, E.; Ryan, C. Towards Incorporating Human Knowledge in Fuzzy Pattern Tree Evolution. In Proceedings of the European Conference on Genetic Programming (Part of EvoStar), Virtual Event, 7–9 April 2021; Springer: Cham, Switzerland, 2021; pp. 66–81.
16. Youssef, A.; Majeed, B.; Ryan, C. Optimizing combinational logic circuits using Grammatical Evolution. In Proceedings of the IEEE 2021 3rd Novel Intelligent and Leading Emerging Sciences Conference (NILES), Giza, Egypt, 23–25 October 2021; pp. 87–92.
17. Majeed, B.; Ryan, C.; McEllin, J.; Youssef, A.; Dias, D.; Murphy, A.; Carvalho, S. Evolving Behavioural Level Sequence Detectors in SystemVerilog Using Grammatical Evolution. In Proceedings of the 15th International Conference on Agents and Artificial Intelligence, Setúbal, Portugal, 22–24 February 2023; Volume 3, pp. 475–483.
18. Mealy, B.; Tappero, F. *Free Range VHDL*; Free Range Factory (2013); eBook: Seattle, Washington, DC, USA, 2018.
19. Chu, P.P. Xilinx Spartan-3 Specific Memory. In *FPGA Prototyping by Verilog Examples*; John Wiley and Sons, Ltd.: Hoboken, NJ, USA, 2008; Chapter 12, pp. 297–308.
20. Ryan, C.; Tetteh, M.K.; Dias, D.M. Behavioural Modelling of Digital Circuits in System Verilog using Grammatical Evolution. In Proceedings of the 12th International Joint Conference on Computational Intelligence, Budapest, Hungary, 2–4 November 2020; SciTePress: Setúbal, Portugal, 2020; pp. 28–39.
21. Bak, W.; Czerniak, J.; Tokarski, T. Intelligent traffic light control system based on multi-agent systems. *Int. J. Comput. Sci. Issues (IJCSI)* **2011**, *8*, 109–115.
22. Yue, S.; Smith, R.K. Applying Context State Machines to Smart Elevators: Design, Implementation and Evaluation. In Proceedings of the 2021 IEEE Symposium Series on Computational Intelligence (SSCI), Orlando, FL, USA, 5–7 December 2021; pp. 1–9. [CrossRef]
23. Siciliano, B.; Khatib, O. *Springer Handbook of Robotics*; Springer: Berlin/Heidelberg, Germany, 2016; pp. 579–619.
24. Stevens, W.R. *TCP/IP Illustrated, Volume 1: The Protocols*; Addison-Wesley: Reading, MA, USA, 1994.
25. Navabi, Z. *VHDL: Modular Design and Synthesis of Cores and Systems*; McGraw-Hill: New York, NY, USA, 2007.
26. Ciletti, M.D. *Advanced Digital Design with the Verilog HDL*, 2nd ed.; Prentice Hall Press: Denver, CO, USA, 2010.
27. Spear, C. *SystemVerilog for Verification, Second Edition: A Guide to Learning the Testbench Language Features*, 2nd ed.; Springer: New York, NY, USA, 2008.
28. Cadence. Cadence Design Systems. 1998. Available online: https://www.cadence.com/ (accessed on 1 April 2022).
29. Morris, M.; Ciletti, M.D. *Digital Design*; Pearson Prentice Hall: Hoboken, NJ, USA, 2007.

30. Fourman, M.P. Compaction of Symbolic Layout Using Genetic Algorithms. In Proceedings of the 1st International Conference on Genetic Algorithms, Pittsburgh, PA, USA, 24–26 July 1985; Lawrence Erlbaum Associates, Inc.: Mahwah, NJ, USA, 1985; pp. 141–153.
31. Cohoon, J.P.; Paris, W.D. Genetic Placement. *IEEE Trans. Comput.-Aided Des. Integr. Circuits Syst.* **1987**, *6*, 956–964. [CrossRef]
32. Kling, R.; Banerjee, P. ESp: Placement by simulated evolution. *IEEE Trans. Comput.-Aided Des. Integr. Circuits Syst.* **1989**, *8*, 245–256. [CrossRef]
33. Higuchi, T.; Niwa, T.; Tanaka, T.; Iba, H.; de Garis, H.; Furuya, T. Evolving Hardware with Genetic Learning: A First Step towards Building a Darwin Machine. In Proceedings of the From Animals to Animats 2: Proceedings of the Second International Conference on Simulation of Adaptive Behavior, Honolulu, HI, USA, 9 August 1993; pp. 417–424.
34. Calvin, W.H. The brain as a Darwin Machine. *Nature* **1987**, *330*, 33–34. [CrossRef] [PubMed]
35. Stomeo, E.; Kalganova, T.; Lambert, C. Generalized disjunction decomposition for evolvable hardware. *IEEE Trans. Syst. Man Cybern. B Cybern.* **2006**, *36*, 1024–1043. [CrossRef] [PubMed]
36. Vasicek, Z.; Sekanina, L. Hardware Accelerators for Cartesian Genetic Programming. In Proceedings of the Genetic Programming, Naples, Italy, 26–28 March 2008; Springer: Berlin/Heidelberg, Germany, 2008; pp. 230–241.
37. Tetteh, M.; Dias, D.M.; Ryan, C. Grammatical Evolution of Complex Digital Circuits in SystemVerilog. *SN Comput. Sci.* **2022**, *3*, 188. [CrossRef] [PubMed]
38. Tetteh, M.K.; Mota Dias, D.; Ryan, C. Evolution of Complex Combinational Logic Circuits Using Grammatical Evolution with SystemVerilog. In Proceedings of the Genetic Programming, Virtual Event, 7–9 April 2021; Hu, T., Lourenço, N., Medvet, E., Eds.; Springer: Cham, Switzerland, 2021; pp. 146–161.
39. Ashlock, D.; Wittrock, A.; Wen, T.J. Training finite state machines to improve PCR primer design. In Proceedings of the 2002 Congress on Evolutionary Computation. CEC'02 (Cat. No.02TH8600), Honolulu, HI, USA, 12–17 May 2002; Volume 1, pp. 13–18. [CrossRef]
40. Klimowicz, A.; Salauyou, V. State Merging and Splitting Strategies for Finite State Machines Implemented in FPGA. *Appl. Sci.* **2022**, *12*, 8134. [CrossRef]
41. Pinto Ferraz Fabbri, S.; Delamaro, M.; Maldonado, J.; Masiero, P. Mutation analysis testing for finite state machines. In Proceedings of the 1994 IEEE International Symposium on Software Reliability Engineering, Monterey, CA, USA, 6–9 November 1994; pp. 220–229. [CrossRef]
42. Villa, T.; Kam, T.; Brayton, R.K.; Sangiovanni-Vincentelli, A.L. *Synthesis of Finite State Machines: Logic Optimization*; Springer Science & Business Media: Berlin/Heidelberg, Germany, 1997.
43. Hemmi, H.; Mizoguchi, J.; Shimohara, K. Development and evolution of hardware behaviors. In Proceedings of the Towards Evolvable Hardware (TEH), Lausanne, Switzerland, 2–3 October 1995; Sanchez, E., Tomassini, M., Eds.; Springer: Berlin/Heidelberg, Germany, 1996; pp. 250–265.
44. Majeed, B.; Carvalho, S.; Dias, D.M.; Youssef, A.; Murphy, A.; Ryan, C. Performance Upgrade of Sequence Detector Evolution Using Grammatical Evolution and Lexicase Parent Selection Method. In Proceedings of the Complex Computational Ecosystems, Baku, Azerbaijan, 25–27 April 2023; Collet, P., Gardashova, L., El Zant, S., Abdulkarimova, U., Eds.; Springer: Cham, Switzerland, 2023; pp. 90–103.
45. Mizoguchi, J.; Hemmi, H.; Shimohara, K. Production genetic algorithms for automated hardware design through an evolutionary process. In Proceedings of the First IEEE Conference on Evolutionary Computation, IEEE World Congress on Computational Intelligence, Orlando, FL, USA, 27–29 June 1994; Volume 2, pp. 661–664.
46. Shanthi, A.; Singaram, L.; Parthasarathi, R. Evolution of asynchronous sequential circuits. In Proceedings of the 2005 NASA/DoD Conference on Evolvable Hardware (EH'05), Washington, DC, USA, 29 June–1 July 2005; pp. 93–96.
47. Soleimani, P.; Sabbaghi-Nadooshan, R.; Mirzakuchaki, S.; Bagheri, M. Using Genetic Algorithm in the Evolutionary Design of Sequential Logic Circuits. *arXiv* **2011**, arXiv:1110.1038. [CrossRef]
48. Ali, B.; Almaini, A.E.A.; Kalganova, T. Evolutionary Algorithms and Theirs Use in the Design of Sequential Logic Circuits. *Genet. Program. Evolvable Mach.* **2004**, *5*, 11–29. [CrossRef]
49. Popa, R.; Aiordăchioaie, D.; Sîrbu, G. Evolvable Hardware in Xilinx Spartan-3 FPGA. In Proceedings of the 2005 WSEAS International Conference on Dynamical Systems and Control (ICDSC), Venice, Italy, 2–4 November 2005; pp. 66–71.
50. Yao, R.; Wang, Y.r.; Yu, S.l.; Gao, G.j. Research on the Online Evaluation Approach for the Digital Evolvable Hardware. In Proceedings of the Evolvable Systems: From Biology to Hardware (ESFBH), Wuhan, China, 21–23 September 2007; Springer: Berlin/Heidelberg, Germany, 2007; pp. 57–66.
51. Xiong, F.; Rafla, N.I. On-chip intrinsic evolution methodology for sequential logic circuit design. In Proceedings of the 2009 52nd IEEE International Midwest Symposium on Circuits and Systems, Cancun, Mexico, 2–5 August 2009; pp. 200–203.
52. Xiong, F.; Tanik, J.; Tanik, M. An Evolutionary Circuit Model for Cardiovascular System: An FPGA Approach. *Int. J. Comput. Inf. Technol. Eng.* **2011**, *5*, 41–53.
53. Tao, Y.; Cao, J.; Zhang, Y.; Lin, J.; Li, M. Using module-level Evolvable Hardware approach in design of sequential logic circuits. In Proceedings of the 2012 IEEE Congress on Evolutionary Computation (CEC), Brisbane, QLD, Australia, 10–15 June 2012; pp. 1–8.
54. Kumar, K. Design of vending machine through implementation of visual automata simulator and finite state machine. *Int. J. Res. Circuits, Devices Syst.* **2021**, *2*, 60–64.

55. Kurniawan, O.; Ismaya, F.; Gata, W.; Septia Nugraha, F.; Lasmana Putra, J. Application of the Finite State Automata Concept in Applications Fruit Vending Machine Simulation. *J. Mantik* **2022**, *6*, 1467–1474.
56. Shivanand, N.; L Rathod, M.; Chetan, S. FPGA based Vending Machine For Logical Gates. In Proceedings of the 2023 3rd International Conference on Smart Data Intelligence (ICSMDI), Trichy, India, 30–31 March 2023; pp. 282–293. [CrossRef]
57. Sindhu, G.; Thanusha, K.; Ganesh, G.V.; Reddy, K.; Pujitha, M. Design of Vending Machine Using Visual Automata Simulator And Finite State Machine. In Proceedings of the 2023 International Conference on Communication, Circuits, and Systems (IC3S), Bhubaneswar, India, 26–28 May 2023; pp. 1–7. [CrossRef]
58. Miller, B.L.; Goldberg, D.E. Genetic Algorithms, Tournament Selection, and the Effects of Noise. *Complex Syst.* **1995**, *9*, 193–212.
59. Spector, L. Assessment of Problem Modality by Differential Performance of Lexicase Selection in Genetic Programming: A Preliminary Report. In Proceedings of the 14th Annual Conference Companion on Genetic and Evolutionary Computation (GECCO '12), Philadelphia, PA, USA, 7–11 July 2012; pp. 401–408.
60. Orzechowski, P.; La Cava, W.; Moore, J.H. Where Are We Now? A Large Benchmark Study of Recent Symbolic Regression Methods. In Proceedings of the Genetic and Evolutionary Computation Conference. Association for Computing Machinery, Kyoto, Japan, 15–19 July 2018; pp. 1183–1190.
61. La Cava, W.; Moore, J. Behavioral search drivers and the role of elitism in soft robotics. In Proceedings of the ALIFE 2018: The 2018 Conference on Artificial Life, Tokyo, Japan, 23–27 July 2018; pp. 206–213.
62. Rodríguez, I.; Rubio, D.; Rubio, F. Complexity of adaptive testing in scenarios defined extensionally. *Front. Comput. Sci.* **2023**, *17*, 173206. [CrossRef]
63. de Lima, A.; Carvalho, S.; Dias, D.M.; Naredo, E.; Sullivan, J.P.; Ryan, C. GRAPE: Grammatical Algorithms in Python for Evolution. *Signals* **2022**, *3*, 642–663. [CrossRef]

algorithms

MDPI

Article

Evolutionary Optimization for the Classification of Small Molecules Regulating the Circadian Rhythm Period: A Reliable Assessment

Antonio Arauzo-Azofra [1], Jose Molina-Baena [2] and Maria Luque-Rodriguez [2,*]

[1] Department of Project Engineering, Universidad de Córdoba, 14014 Córdoba, Spain; arauzo@uco.es
[2] Department of Computer Science and Artificial Ingelligence, Universidad de Córdoba, 14014 Córdoba, Spain; i42mobaj@uco.es
* Correspondence: mluque@uco.es

Abstract: The circadian rhythm plays a crucial role in regulating biological processes, and its disruption is linked to various health issues. Identifying small molecules that influence the circadian period is essential for developing targeted therapies. This study explores the use of evolutionary optimization techniques to enhance the classification of these molecules. We applied a genetic algorithm to optimize feature selection and classification performance. Several tree-based learning classification algorithms (Decision Trees, Extra Trees, Random Forest, XGBoost) and a distance-based classifier (*k*NN) were employed. Their performance was evaluated using accuracy and F1-score, while considering their generalization ability with a validation set. The findings demonstrate that the proposed genetic algorithm improves classification accuracy and reduces overfitting compared to baseline models. Additionally, the use of variance in accuracy as a penalty factor may enhance the model's reliability for real-world applications. Our study confirms that evolutionary optimization is an effective strategy for classifying small molecules regulating the circadian rhythm. The proposed approach not only improves predictive performance but also ensures a more robust model.

Keywords: feature selection; genetic algorithm; high dimensionality; few instances

1. Introduction

The circadian rhythm regulates biological processes in 24 h cycles, and its alteration is associated with sleep disorders and can be related to pathologies such as cancer [1]. The circadian clock plays a fundamental role in the regulation of sleep, and its stability is increasingly recognized as a determining factor in various biological processes, which are essential for maintaining good health. To this end, photopharmacological manipulation of a core clock protein, mammalian Cryptochrome 1 (CRY1), is an effective strategy for the regulation of the circadian clock. CRY1 is a protein that influences the circadian period. Supported by the analysis of data and advanced computational modelling techniques, it has been established as an essential tool in the detection and characterization of compounds with an impact on circadian dynamics [2]. Recent studies have applied high-throughput screening and data analysis to identify potential drugs, taking advantage of advanced modelling and machine learning techniques [3].

Structural approaches in drug discovery optimize efficiency and reduce costs, while the incorporation of screening methods allows the elimination of inappropriate molecules, such as toxic or inactive ones. After identifying 171 molecules that target functional domains of the CRY1 protein, using structure-based drug design methods, and experimentally determining that 115 of these molecules were nontoxic, Gul et al. [4] performed a machine learning study to classify molecules by identifying features that make them toxic. They also addressed the classification of the same molecules based on their effect on CRY1. While both problems are considered challenging, only the second has been further investigated in other studies [5].

The problem of learning to determine toxicity from these 171 molecules is complex, characterized by many features and few examples. As a result, it is considered ill-posed, as traditional statistical methods often struggle with insufficient data. In this context, overfitting becomes a significant risk, as models can easily memorize the training data instead of generalizing. Therefore, we classify it as an overfitting-prone problem. Tackling these handicaps constitutes an interesting challenge in machine learning [6]. These handicaps motivate us to apply improved automated feature selection while remaining fully aware of its limitations.

In this paper, we reproduce the machine learning experimentation from [4], revealing its weaknesses and evaluating the use of a more advanced feature selection process based on genetic algorithm meta-heuristic search to achieve improved and more trustworthy results.

The main contributions of this paper are as follows:

- Demonstration of the need for validation: We make evident the importance of incorporating validation after training and feature selection to avoid overfitting. In addition, we provide reliable estimates of the best performance achieved with this dataset.
- Genetic algorithm for automated feature selection: We propose a genetic algorithm (GA)-based framework for automated feature selection, which achieves results comparable to or better than manual Recursive Feature Elimination (RFE), reducing human effort and bias.
- Test variance as a generalization criterion: We propose the use of variance across cross-validation folds as an additional objective during feature selection, promoting models that not only perform well but also generalize better across different data splits.

2. Genetic Algorithm for Feature Selection

Evolutionary algorithms, inspired by natural evolution, are designed to optimize solutions. Among them, genetic algorithms (GAs) are particularly well-suited for feature selection (FS). These algorithms start with a population of individuals, each representing a possible solution. Through processes of selection, crossover, and mutation, successive generations evolve towards better solutions, until the number of generations specified by parameter N is reached (Figure 1). The following subsections describe the main aspects of the algorithm used in this work.

2.1. Evolution

The framework of the algorithm is based on the evolution of a single population of solutions over multiple generations.

Once the population has been initialized, each evolutionary cycle begins with the application of crossover and mutation operators to generate new solutions. Subsequently, the individuals in the population are evaluated and selected, based on their performance in the target task, to form the next generation with the objective of progressively improving the quality of the solutions. Figure 1 illustrates the main steps of the algorithm.

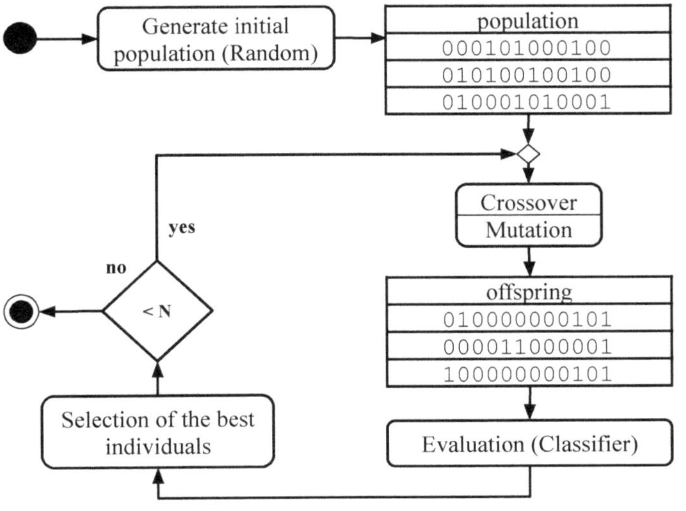

Figure 1. Genetic algorithm activity diagram.

2.2. Encoding

The encoding of solutions determines how potential solutions to the problem are represented. Each solution is encoded as a genotype, which serves as an abstract representation of the solution. To evaluate its quality, the genotype is decoded into its corresponding phenotype, representing the actual interpretation within the problem domain. The effectiveness of the encoding scheme directly influences the algorithm's ability to explore and optimize solutions efficiently.

In this work, the chosen encoding scheme is binary. Each individual in the population represents a possible solution to the problem, which is described as a vector of length n. Each value in the vector corresponds to a feature of the problem, with a 1 indicating that the feature is selected and a 0 otherwise.

2.3. Initialization

In an evolutionary algorithm, the initial population is generated randomly, assigning to each individual an initial configuration that represents a possible solution to the problem. However, the nature of the problem significantly influences the optimal configuration of individuals. Depending on the dimensionality of the dataset and the complexity of the problem, the expected number of selected features may vary considerably.

GAs typically initialize individuals with a uniform probability of inclusion or exclusion of each feature, resulting in an expected selection of 50% of the features. However, in high-dimensionality datasets, this strategy can generate overfitted initial solutions, which significantly slows down the evaluation of individuals. Since the GA will naturally adjust the number of features selected based on their impact on performance, a more efficient initialization strategy may be to start with a smaller number of features when working with high-dimensional spaces [7].

To strike a balance between simplicity and generality, the initialization strategy used in our algorithm is based on an adjustable likelihood as parameter [8].

2.4. Crossover Operator

The crossover operation combines coded individuals to explore potentially better solutions. In this work, two-point crossover, a method widely used in GAs, is employed to generate the next generation within the population. This mechanism exchanges a segment of the genotypic vector between two parent individuals, bounded by two randomly selected points. Figure 2 illustrates this process. The probability of selecting segments of different sizes and their location within the genotype is uniform, ensuring a balanced exploration of the search space without introducing biases in the optimization.

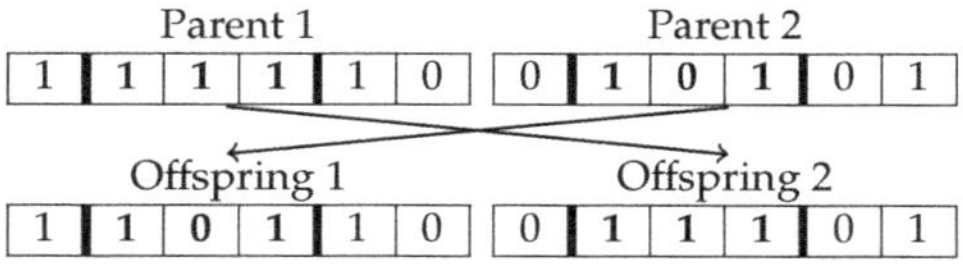

Figure 2. Crossover operator.

2.5. Mutation Operator

The mutation operation on encoded individuals introduces random modifications in the population to preserve genetic diversity. This mechanism is crucial for preventing premature convergence, allowing the algorithm to explore a broader search space and reducing the risk of getting trapped in local optima. In this study, the bit-flip mutation operator is employed, a widely used technique in binary-encoded GAs. This operator selectively alters certain binary values within the genotype of each individual, flipping them with a predefined probability, which serves as a control parameter for the mutation intensity. Figure 3 illustrates this process.

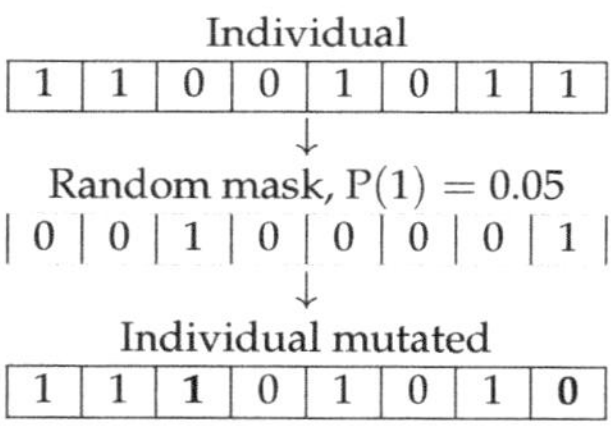

Figure 3. Mutation operator.

2.6. Fitness Evaluation

Fitness evaluation in a GA is the process of assessing how well each individual in the population solves the given problem. The fitness function quantitatively measures the quality of each solution based on a predefined criterion. In the case of feature selection problems, this criterion is the efficiency of the selected feature subset. The fitness score determines the likelihood of an individual being selected for reproduction, guiding the evolutionary process toward optimal solutions over successive generations.

In this work, we use the wrapper approach with 10-fold cross-validation to evaluate the quality of each individual (feature subset). The fitness of an individual is determined by training a machine learning model using the selected features, represented by the genotype, and evaluating its performance based on the accuracy metric. Furthermore, to encourage compact feature subsets, we include a penalty term to balance performance and feature reduction (Equation (1)).

$$\text{Fitness} = (1-\alpha)\cdot\text{Effectiveness} + \alpha\cdot\frac{|F|-|S|}{|F|} \tag{1}$$

where α is the weight parameter, $|S|$ is the number of selected features, $|F|$ is the total number of features in the dataset, and Effectiveness is the the average accuracy across all test folds.

To mitigate overfitting, we define a second fitness function (Equation (2)), which incorporates the variance of accuracy as an additional penalty factor. The smaller the variance in accuracy, the more stable the model's behavior, meaning it performs consistently across all test runs. A lower variance indicates that the model generalizes well, reducing the likelihood of overfitting. This stability suggests that the algorithm is not overly dependent on specific training data patterns. Our hypothesis is that this will make it a better generalizer, more reliable for real-world applications.

$$\text{Fitness} = (1-\alpha)\cdot(\text{Effectiveness} - \text{variance}) + \alpha \cdot \frac{|F|-|S|}{|F|} \quad (2)$$

2.7. Selection

Selection in a GA is the process of choosing individuals from the current population to create offspring for the next generation. The selection method directly influences the convergence and performance of the algorithm by favoring individuals with higher fitness scores while maintaining genetic diversity. The goal of selection is to balance exploration and exploitation, ensuring that the algorithm effectively searches the solution space while avoiding premature convergence.

In this work, we employ a binary tournament selection approach, where two individuals are randomly selected from the population, and the one with the higher fitness value is chosen as a parent for the next generation. This process is repeated until the required number of parents is selected. Binary tournament selection balances exploration and exploitation of the search space, allowing the most promising solutions to propagate across generations while preserving genetic diversity.

3. Differential Evolution for Feature Selection

Another evolutionary algorithm suitable for feature selection is Differential Evolution (DE) adapted for discrete or binary domains. DE is a population-based stochastic optimization algorithm that has demonstrated strong performance on complex, nonlinear, and high-dimensional problems [9]. Its general procedure is outlined in Algorithm 1.

Algorithm 1 Differential Evolution (DE)

Initialize population $P = \{x_1, x_2, \ldots, x_{NP}\}$ with random values
Evaluate fitness $f(x_i)$ for all $x_i \in P$
for $g = 1$ to $G_{\max}$ **do**
  for each individual x_i in P **do**
    Randomly select distinct $x_a, x_b, x_c \in P$ such that $i \neq a \neq b \neq c$
    for $j = 1$ to D **do**
      $v_{ij} = x_{aj} + F \cdot (x_{bj} - x_{cj})$
      Generate $r \sim \mathcal{U}(0,1)$
      if $r < CR$ **then**
        $u_{ij} = v_{ij}$
      else
        $u_{ij} = x_{ij}$
      end if
    end for
    if $f(u_i) \leq f(x_i)$ **then**
      $x_i \leftarrow u_i$
    end if
  end for
end for
return Best individual in P

Originally designed for continuous parameter optimization, DE has been successfully adapted to feature selection tasks by encoding candidate solutions as binary vectors that represent feature subsets [10]. In the context of feature selection, each binary vector represents a subset of features and the objective is to optimize the fitness function f (wrapper classification accuracy). DE offers a flexible approach for feature selection with two control parameters (F and CR). It evolves the population over successive generations using three main operators:

- Mutation: Weighted difference of two randomly selected population vectors (x_b, x_c) with a third one (x_a). This binary version [10] uses XOR as the difference operator and the F weighting parameter modulated by a random choice as the product and addition operators. This avoids bias on the appearance of 0 and 1 values.

$$v_i = x_a + \mathrm{F} \cdot (x_b - x_c) \tag{3}$$

- Crossover: To increase diversity, elements from the mutant vector (v_i) are combined with the target vector with CR probability to create a trial vector.

$$u_{ij} = \begin{cases} v_{ij} & \text{if } \text{rand}_j \leq CR \\ x_{ij} & \text{otherwise} \end{cases} \tag{4}$$

- Selection: The fitness of the trial vector is evaluated, and it replaces the target vector in the next generation if it yields a better fitness value.

4. Material and Methods

This section describes the experimental methodology used to reproduce the previous experimentation and to evaluate the performance of feature selection with a GA metaheuristic search on the Toxicity classification problem. In this study, we aimed to validate three main hypotheses. First, we hypothesize that the proposed validation method will demonstrate the presence of overfitting in the results reported by the previous study. Second, we aim to improve the results of the compared approach by leveraging an GA. Third, the use of variance of accuracy as a penalty factor will make it a better generalizer. The experiments were designed to assess the effectiveness of our approach under different conditions. We provide details on the dataset characteristics, the classifiers used in the experiments, the execution environment, and the parameters adopted in the GA.

4.1. Dataset

The dataset comes from the research by Gul et al. [4]. Two problems are addressed in that research and we focused on the first one, detecting toxicity by molecular descriptors. The dataset is available in the UCI repository [11] under the name Toxicity.

The Toxicity dataset was developed to evaluate the toxicity of molecules designed to interact with CRY1, a protein central to the regulation of the circadian rhythm. This biological clock influences numerous physiological processes and its disruption has been associated with diseases such as cancer and metabolic disorders.

The dataset (Table 1) contains molecular descriptors for 171 molecules obtained by computational calculations, which can be used to train machine learning models capable of predicting whether a molecule is toxic or non-toxic. Each molecule in the dataset is represented by 1203 molecular descriptors, which include physicochemical, topological and structural properties. Examples of descriptors include the following:

- Physicochemical properties: Molecular mass, logP (partition coefficient), number of hydrogen bonds.

- Topological descriptors: molecular connectivity indices, number of cycles in the structure.
- Electronic properties: Energy of orbitals, electrostatic potential.

These descriptors are generated by computational chemistry software and are commonly used in toxicity prediction models.

Non-toxic, the majority class, accounts for 67.25% of the instances. Therefore, a classifier that always predicts the majority class would achieve an accuracy of 67.25%. As a result, a good performance from any algorithm should exceed this baseline to demonstrate its effectiveness in distinguishing between classes.

Table 1. Dataset properties.

Features (Molecular Descriptors)	Instances (Molecules)	Clases	Class Distribution Non-Toxic	Toxic
1203	171	2	115	56

4.2. Classifiers

The DTC, RFC, ETC, and XGBC classifiers used in [4] and *k*NNwere employed for the experimentation. While the former are all tree-based ensemble methods, *k*NN was included as a non-tree-based, instance-based learning algorithm to provide a comparative perspective and evaluate its performance under the same feature selection strategies.

*k*NN (*k*-Nearest Neighbors) is a nonparametric instance-based learning algorithm used for classification and regression tasks [12,13]. It classifies a new instance by considering the K closest training examples in the feature space, typically measured using Euclidean distance or other distance metrics. The predicted class is determined by a majority vote among the nearest neighbors.

DTC (Decision Tree Classifier) is a nonparametric supervised learning algorithm, which is used for both classification and regression tasks [14]. It has a hierarchical tree structure, consisting of a root node, branches, internal nodes and leaf nodes.

RFC (Random Forest Classifier) is a classifier that relies on combining a large number of uncorrelated and weak decision trees to arrive at a single result [15].

ETC (Extra Trees Classifier) is a classification algorithm based on decision trees, similar to RFC, but with more randomization [16]. Instead of searching for the best splits at each node, it randomly selects features and cutoff values. This makes it faster and less prone to overfitting. It works by creating a set of decision trees and makes predictions by majority vote.

XGBC (Extreme Gradient Boosting Clasiffier) is a machine learning algorithm based on boosting, which builds sequential decision trees to correct errors in previous trees [17]. It is efficient, fast, and avoids overfitting thanks to its regularization. It uses gradient descent to optimize the model and is known for its high classification accuracy. However, it can be complex to fit and has less interpretability than other models.

4.3. Development and Running Environment

The GA for feature selection was programmed in Python, using the library DEAP (Distributed Evolutionary Algorithms in Python), an evolutionary computation framework for rapid prototyping and testing of ideas [18]. It seeks to make algorithms explicit and data structures transparent. Likewise, we used Scikit-learn [19], which is a free software machine learning library for the Python programming language. Experiments were run on a cluster of 6 nodes with Intel Xeon E5420 CPU 2.50GHz processor, under an Ubuntu 22.04 GNU/Linux operating system.

4.4. Experimental Parameters

The experimental setup initially attempted to mimic the study with which we compared ours, the original study that defined the dataset [4]. The same grid search using train–test 10 fold cross-validation was applied to choose the parameter values from those described in Table 2. This included the parameters for the *k*NN classifier, which was incorporated into our experimentation to introduce diversity by considering a proximity-based classifier.

Table 2. Parameters of the classifiers tested in grid search.

Parameter †	*k*NN	DTC	RFC and ETC	XGBC
k	[1,3,5,7,9]	—	—	—
max_depth	—	[1, 2, ... , 10, None]	[1, 2, ... , 6, None]	[3, 5, 7, 10]
min_samples_splits	—	[2, 3, ... , 10]	[2–5]	—
min_samples_leafs	—	[1, 2, ... , 10]	[1–5]	—
max_features	—	[1, 2, ... , num_features]	[1, 2, ... , sqrt(num_features)]	—
n_estimators	—	—	[100, 200]	[100, 200]
learning_rate	—	—	—	[0.01, 0.1]
min_child_weight	—	—	—	[1, 3, 5]
subsample	—	—	—	[0.5, 0.7]
colsample_bytree	—	—	—	[0.5, 0.7]

† **k:** The number of nearest neighbors used to classify in *k*NN. **max_depth:** The maximum depth of the tree. **min_samples_splits:** The minimum number of samples required to split an internal node. **min_samples_leafs:** The minimum number of samples required to be at a leaf node. **max_features:** The number of features to consider when looking for the best split. **n_estimators:** The number of trees in the forest. **learning_rate:** Step size shrinkage used in update to prevent overfitting. **min_child_weight:** Minimum sum of instance weight needed in a child. **subsample:** Subsample ratio of the training instances. **colsample_bytree:** Subsample ratio of columns when constructing each tree.

To evaluate the performance of the classification, the same internal cross-validation of ten partitions from grid search optimization was used in the original study. To obtain stable results, they repeated the experimentation 100 times. However, given that the number of combinations evaluated by grid search is in the order of the tens of thousands, overfitting is likely to occur, potentially leading to an overestimation of the expected accuracy. To address this issue, instead of repeating the experimentation 100 times, we utilized nested cross-validation with 100 folds. In this way, the results were expected to be similar in testing, as only a very small number of instances were omitted in each run, while providing a reliable estimate of the expected performance. The whole process was repeated 10 times to ensure a stable final validation result. This evaluation is illustrated in Figure 4.

After reproducing the experimentation with the 13 features selected in the original study, the GA was applied with the goal of finding a better feature set fitting more effective classifiers. In order to configure the parameters of the GA, based on previous experimentation, we adopted the values proposed in [20], which are detailed in Table 3. Since using grid search inside the wrapper cross validation (used as fitness measure) is unfeasible, the default parameters from scikit-learn were used for the classifiers and 1 nearest neighbor for *k*NN.

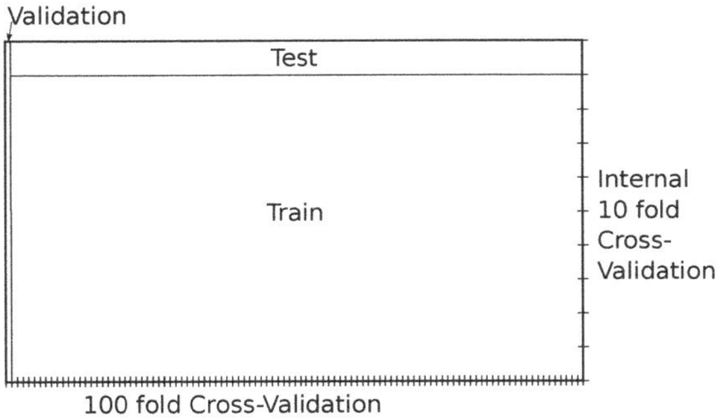

Figure 4. Illustration of one iteration of the nested cross-validation.

Table 3. Parameter values fixed in the genetic algorithm.

Parameter	Description	Value
p_crossover	Crossover probability	0.75
p_mutation	Mutation probability	0.15
elitims	Number of best individuals kept for next generation	0
init_prob	Probability initial features	0.01
population	Number of individuals in each population	50
N	Number of generations to evolve	300

In order to find a good setup for feature selection GA, the parameter values shown in Table 4 were tested.

Table 4. Parameter values tested for the genetic algorithm.

Parameter	Description	Values
penalty	Weight for reduction goal in fitness	0.3, 0.5, 0.7
var_penalty	Variance penalty applied	Yes, No

As part of our comparative analysis, we also executed DE. Its parameter settings are summarized in Table 5. The control parameters F and CR were selected from commonly used default values in the DE literature, based on preliminary experiments on other datasets. The initialization strategy was aligned with that of the genetic algorithm (GA), after observing that the typical default initialization value of 0.5 yielded inferior results during preliminary testing. To ensure a comparable computational effort, the same population size and number of generations were used for both algorithms.

Table 5. Parameter values for Differential Evolution.

Parameter	Description	Value
CR	Crossover probability	0.9
F	Weight for difference acceptance	0.25
init_prob	Probability initial features	0.01
population	Number of individuals in each population	50
G_{max}	Number of generations to evolve	300

5. Results Analysis

As an initial step, we replicated the experimental evaluation presented in [4], using the same 13 selected features and sticking to the same classification processes and software for all classifiers, including the *k*NN classifier. This replication ensures methodological consistency and enables a direct comparative analysis with the original results. The only difference is that we transformed the 100 repetitions into a second-level 100-fold cross-validation by prescinding one or two instances per repetition. Given the small number of omitted instances, their impact on the results is expected to be negligible, while it establishes a confident approach to evaluate potential improvements and assess the robustness of the models.

The results achieved using Recursive Feature Elimination (RFE) for each classifier in the original experimentation are shown in the first column of Table 6. Then, they performed a second RFE and selected 13 features as the most relevant. This is shown in the second column and it is the part of the experimentation reproduced. The classification results obtained with this subset of features in the reproduced experimentation are shown in the following columns. It is unclear to us what the value reported in the original proposal is. We believe it must be the average test accuracy of the best model found, not the average of the 100 runs over the average test, because that value is closer to the best model column in Table 6 (which is the best model according to the average of 100 repetitions of 10-fold CV test accuracies). With that consideration, the results obtained seem similar, confirming the validity of their experimentation.

Table 6. Results from [4] compared with the models found in the experimental reproduction with 100 repetitions of the 10-fold CV using the same 13 features.

	Original [4]		**100 Repetitions, Test Accuracy**			**10 × 100 CV**
Classifier	**Best (Feat)**	**13 Feat.**	**Worst**	**Average ± Std**	**Best**	**Validation**
*k*NN	—	—	0.6276	0.6384 ± 0.0066	0.6570	0.6427
DTC	0.7877 (19)	0.7963	0.7217	0.7479 ± 0.0121	0.7765	0.7409
RFC	0.7299 (9)	0.7236 †	0.6765	0.6921 ± 0.0077	0.7158	0.6620
ETC	0.7136 (20)	0.6857 †	0.6688	0.6764 ± 0.0064	0.6941	0.6374
XGBC	0.7117 (17)	0.6887 †	0.6688	0.6805 ± 0.0057	0.7040	0.6404

Values in green indicate validation accuracy over the majority rate. † Best result reported with 13 features but not necessarily the same 13 features identified as best for DTC.

Nevertheless, it is important to note that validation indicates that the model's expected accuracy is around 74%, which is lower than the reported best test average of 79.63% from the original study (and the 77.65% from the reproduced equivalent result). It seems that, by repeating the train–test process many times, the feature selection and the parameters of the classifiers may be overfit to the test data.

The DTC classifier achieved the highest validation accuracy (74.09%), demonstrating its effectiveness when trained with the selected features. However, the performance drop observed for the remaining classifiers between testing and validation suggests that the selected characteristics may primarily favor the DTC, potentially limiting its generalization across different models. Furthermore, it is important to note that the performance of the other classifiers does not exceed the baseline of the majority class, indicating that these models provide very poor discriminative power. To highlight this, those models with validation accuracy over the majority rate are colored in green. These results highlight the need for further analysis of the feature selection and learning process to ensure robustness and generalization.

In the second phase of our study, the GA is used to perform feature selection and evaluate its impact on the different classifiers. As an example, the evolution process of one run of the GA is illustrated in Figure 5.

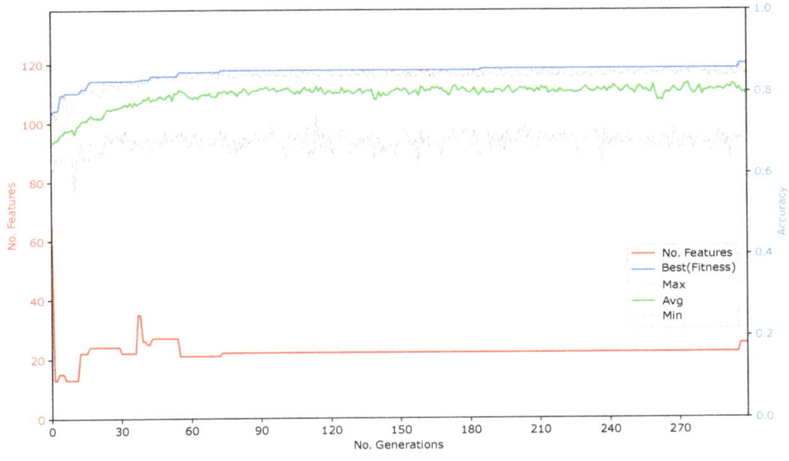

Figure 5. Evolution of the GA wrapping DTC with a 0.3 feature set penalty and no variance penalty (fourth line of Table 7).

The results in Table 7 show that the 16 features selected by the GA wrapping RFC achieved the highest validation accuracy (71.64%) by using a penalty factor of 0.3, demonstrating that the evolutionary approach can improve the generalization capacity of the models studied, as this result is much higher than those achieved for RFC by using RFE in the previous study. In contrast, the DTC classifier, which had previously obtained the highest test accuracy, experienced a significant drop in validation, evidencing a possible overfitting of the training set, probably because it is using too many features.

In the third phase, we test the hypothesis that when using Equation (2), the found feature sets are less prone to overfitting. Table 8 shows the results of the same experiments presented in Table 7 but using the new fitness calculation with a penalization for non homogeneous accuracy among test partitions. The results seem to improve but the differences are small and they cannot be considered conclusive. For this reason, we think that more research is needed to find strategies to avoid overfitting in these challenging high-dimensional datasets with few instances.

Table 7. Results using features selected by the genetic algorithm.

Classifier	noF	Penalty	Min.	Avg. Test Accuracy ± Std	Max.	Validation
*k*NN	10	0.3	0.6456	0.6596 ± 0.0091	0.6868	0.6702
	11	0.5	0.6269	0.6519 ± 0.0091	0.6750	0.6117
	8	0.7	0.6397	0.6606 ± 0.0095	0.6868	0.6550
DTC	25	0.3	0.7283	0.7478 ± 0.0100	0.7757	0.6444
	25	0.5	0.7224	0.7423 ± 0.0099	0.7699	0.6690
	12	0.7	0.7235	0.7411 ± 0.0082	0.7588	0.6538
RFC	16	0.3	0.7158	0.7359 ± 0.0090	0.7577	0.7164
	9	0.5	0.6989	0.7225 ± 0.0085	0.7471	0.6942
	7	0.7	0.7048	0.7214 ± 0.0074	0.7460	0.7047
ETC	18	0.3	0.6868	0.7051 ± 0.0103	0.7279	0.7018
	12	0.5	0.6706	0.6902 ± 0.0114	0.7165	0.6784
	9	0.7	0.6746	0.6905 ± 0.0079	0.7059	0.6579
XGBC	14	0.3	0.7044	0.7216 ± 0.0089	0.7574	0.6690
	9	0.5	0.7235	0.7447 ± 0.0117	0.7813	0.7035
	9	0.7	0.7176	0.7356 ± 0.0100	0.7640	0.6965

Values in green indicate validation accuracy over the majority rate.

Table 8. Results using features selected by the genetic algorithm with a fitness penalization based on the variance in CV.

Classifier	noF	Penalty	Min.	Avg. Test Accuracy ± Std	Max.	Validation
kNN	10	0.3	0.6397	0.6598 ± 0.0091	0.6868	0.6702
	9	0.5	0.6393	0.6616 ± 0.0099	0.6868	0.6714
	4	0.7	0.6564	0.6776 ± 0.0089	0.6985	0.7070
DTC	25	0.3	0.7283	0.7593 ± 0.0119	0.7930	0.6871
	18	0.5	0.7401	0.7662 ± 0.0091	0.7930	0.6830
	15	0.7	0.7404	0.7661 ± 0.0108	0.7996	0.7228
RFC	19	0.3	0.7099	0.7330 ± 0.0094	0.7577	0.7245
	9	0.5	0.7169	0.7329 ± 0.0071	0.7522	0.7170
	7	0.7	0.6824	0.7030 ± 0.0103	0.7279	0.6661
ETC	37	0.3	0.6688	0.6758 ± 0.0047	0.6941	0.6456
	13	0.5	0.6765	0.6963 ± 0.0086	0.7221	0.6906
	11	0.7	0.6746	0.6927 ± 0.0080	0.7099	0.6743
XGBC	34	0.3	0.7048	0.7181 ± 0.0086	0.7518	0.6930
	13	0.5	0.6864	0.7047 ± 0.0085	0.7335	0.6906
	9	0.7	0.6981	0.7162 ± 0.0092	0.7529	0.6848

Values in green indicate validation accuracy over the majority rate.

Table 9 compares the different methods analyzed, including RFE, the GA approach, the variance penalty version of the GA, and DE applied to the five classifiers kNN, DTC, RFC, ETC, and XGBC. The results show that with RFE, DTC achieved the best validation accuracy (74.09%), separating itself from the other classifiers. However, in the GA and GA with variance penalty methods, RFC and XGBC showed superior performance, especially with penalties of 0.3 and 0.5. In particular, RFC achieved a remarkable validation performance (71.64%) with 16 features selected using the GA approach, while XGBC reached its maximum performance (70.35%) with 9 features and a penalty of 0.5. However, kNN showed inferior performance in most cases, although a higher penalty (0.7) improved its accuracy by selecting only four features. These results suggest that GA and GA with variance penalty methods are effective in improving generalization, especially in ensemble classifiers such as RFC and XGBC, while DTC remains the best classifier with the 13 features selected in the original study. With a similar running time, DE have not improved any of the results from RFE or GA.

Table 9. Comparison of feature selection for each classifier based on validation results.

Type	P.	kNN		DTC		RFC		ETC		XGBC	
		noF	Val.	noF	Val.	noF	Val.	noF	Val.	noF	Val.
RFE [4]	—	13	0.6427	13	**0.7409**	13	0.6620	13	0.6374	13	0.6404
GA	0.3	10	0.6702	25	0.6444	16	0.7164	18	**0.7018**	14	0.6690
	0.5	11	0.6117	25	0.6690	9	0.6942	12	0.6784	9	**0.7035**
	0.7	8	0.6550	12	0.6538	7	0.7047	9	0.6579	9	0.6965
GA with VarPenalty	0.3	10	0.6702	25	0.6871	19	**0.7245**	37	0.6456	34	0.6930
	0.5	9	0.6714	18	0.6830	9	0.7170	13	0.6906	13	0.6906
	0.7	4	**0.7070**	15	0.7228	7	0.6661	11	0.6743	9	0.6848
DE	—	14	0.5883	33	0.6035	72	0.6708	41	0.6719	14	0.6491

Values in green indicate validation accuracy over the majority rate. Boldfaced values are the best results for each classifier.

Although accuracy was used as the primary metric for comparison, the F1-score was also calculated to provide another assessment of model performance, given that this dataset has moderate class imbalance, which is common in medical datasets. This additional metric allows for an assessment of the balance between precision and recall in the classifiers evaluated [21]. Therefore, its inclusion complements the accuracy metric by highlighting the trade-off between false positives and false negatives, which is especially relevant in medical decision-making contexts.

F1-score (F_1) is defined as the harmonic mean of precision and recall, providing a single measure that balances both concerns, as shown in Equation (5),

$$F_1 = 2 \cdot \frac{Precision \cdot Recall}{Precision + Recall} \tag{5}$$

where *Precision* is the ratio of true positive predictions to the total number of positive predictions made, as shown in Equation (6),

$$Precision = \frac{\text{True Positives}}{\text{True Positives} + \text{False Positives}} \tag{6}$$

and *Recall* is the ratio of true positive predictions to the total number of actual positive instances, as shown in Equation (7).

$$Recall = \frac{\text{True Positives}}{\text{True Positives} + \text{False Negatives}} \tag{7}$$

F_1 ranges from 0 to 1, where 1 indicates perfect precision and recall and 0 indicates the worst performance. As we considered Toxic as the positive class, in this metric, the baseline established by majority classifiers, that we set to be beaten, is given by Equation (8) or (9), where p is the probability of the majoritarian class (0.6725).

$$\text{If negative class is majoritarian:} \quad F_1 = 0 \tag{8}$$

$$\text{If positive class is majoritarian:} \quad F_1 = \frac{2p}{1+p}, \quad \text{where } p > 0.5 \tag{9}$$

As in this case the majoritarian class is the negative class (Non-toxic), any classifier with F_1 over 0 is performing better than a majority classifier with respect to detecting toxicity.

The results in Table 10 show that the best performing classifiers coincide with those identified by the accuracy metric. The values show that there is much space to improve in this challenging dataset.

Table 10. Comparison of feature selection methods for each classifier based on validation results using the F1-score metric.

Type	P.	*k*NN		DTC		RFC		ETC		XGBC	
		noF	Val.	noF	Val.	noF	Val.	noF	Val.	noF	Val.
RFE [4]	—	13	0.5772	13	**0.7088**	13	0.5717	13	0.5437	13	0.5433
GA	0.3	10	0.6259	25	0.5920	16	0.6435	18	**0.6248**	14	0.6057
	0.5	11	0.5540	25	0.6142	9	0.6281	12	0.6033	9	**0.6417**
	0.7	8	0.5994	12	0.6049	7	0.6287	9	0.5715	9	0.6404
GA with VarPenalty	0.3	10	0.6271	25	0.6384	19	**0.6505**	37	0.5456	34	0.6367
	0.5	9	0.6273	18	0.6313	9	0.6491	13	0.6168	13	0.6329
	0.7	4	**0.6612**	15	0.6747	7	0.5853	11	0.5951	9	0.6214
DE	—	14	0.5148	33	0.5238	72	0.5618	41	0.5630	14	0.5495

Boldfaced values are the best results for each classifier.

6. Conclusions

The problem addressed is highly challenging, not only because it belongs to the class of high-dimensional datasets with few instances but also because, after extensive experimentation, achieving a genuine improvement in generalization over the majority class rate with high confidence appears to be highly unlikely.

After reproducing the experiments from the paper that introduced the problem, using an independent validation, we found that the actual expected accuracy is 4% lower than the reported test accuracy. This highlights the importance of avoiding repeated testing on the same data, as it increases the likelihood of obtaining a model that performs well under a specific test setup but fails to generalize.

Using the proposed GA to automate the FS process appears promising as, although it has not been able to improve the best model found using DTC in the original study, it has improved the FS performed with RFE for most of the classification models.

The use of the proposed variance penalty in the GA's fitness function seems promising because it has achieved several better generalization results than the non-penalized version. However, it deserves more research to finetune it and prove its performance in different datasets.

Author Contributions: Conceptualization, M.L.-R., J.M.-B. and A.A.-A.; methodology, J.M.-B.; software, J.M.-B. and A.A.-A.; validation, A.A.-A., M.L.-R. and J.M.-B.; formal analysis, M.L.-R.; investigation, J.M.-B. and A.A.-A.; resources, J.M.-B.; data curation, J.M.-B. and A.A.-A.; writing—original draft preparation, J.M.-B. and A.A.-A.; writing—review and editing, M.L.-R. and A.A.-A.; supervision, M.L.-R. and A.A.-A.; project administration, M.L.-R.; funding acquisition, A.A.-A. All authors have read and agreed to the published version of the manuscript.

Funding: This research is supported by projects PID2020-118224RB-I00, PID2023-151336OB-I00 and PID2023-148396NB-I00 funded by MICIU/AEI/10.13039/501100011033, Ministerio de Ciencia e Innovación (Spain, EU).

Data Availability Statement: The dataset used is available at: https://archive.ics.uci.edu/dataset/728/toxicity-2. Source code is available at: https://github.com/arauzo/EvolOptimizationClassification-Small-Molecules-Regulating-the-Circadian-Rhythm-Period. Experimental logs are available upon request to authors.

Conflicts of Interest: The authors declare no conflicts of interest.

Abbreviations

The following abbreviations are used in this manuscript:

CV	Cross Validation
CRY1	Cryptochrome 1
DEAP	Distributed Evolutionary Algorithms in Python
DE	Differential Evolution
DTC	Decision Tree Classifier
ETC	Extra Trees Classifier
FS	Feature Selection
GA	Genetic Algorithm
kNN	k-Nearest Neighbors
RFC	Random Forest Classifier
RFE	Recursive Feature Elimination
XGBC	Extreme Gradient Boosting Classifier

References

1. Jagielo, A.; Benedict, C.; Spiegel, D. Circadian, hormonal, and sleep rhythms: Effects on cancer progression implications for treatment. *Front. Oncol.* **2023**, *13*, 1269378. [CrossRef] [PubMed]
2. Kolarski, D.; Miller, S.; Oshima, T.; Nagai, Y.; Aoki, Y.; Kobauri, P.; Srivastava, A.; Sugiyama, A.; Amaike, K.; Sato, A. Others Photopharmacological manipulation of mammalian CRY1 for regulation of the circadian clock. *J. Am. Chem. Soc.* **2021**, *143*, 2078–2087. [CrossRef] [PubMed]
3. Wildey, M.; Haunso, A.; Tudor, M.; Webb, M.; Connick, J. High-throughput screening. *Annu. Rep. Med. Chem.* **2017**, *50*, 149–195.
4. Gul, S.; Rahim, F.; Isin, S.; Yilmaz, F.; Ozturk, N.; Turkay, M.; Kavakli, I. Structure-based design and classifications of small molecules regulating the circadian rhythm period. *Sci. Rep.* **2021**, *11*, 18510. [CrossRef] [PubMed]
5. Chawathe, S. Attribute Selection and Visualization for Classification for Small Molecules. In Proceedings of the 2022 IEEE 13th Annual Ubiquitous Computing, Electronics and Mobile Communication Conference (UEMCON), New York, NY, USA, 26–29 October 2022; pp. 0456–0462.
6. Kuncheva, L.; Matthews, C.; Arnaiz-González, Á; Rodrıéguez, J. Feature selection from high-dimensional data with very low sample size: A cautionary tale. *arXiv* **2020**, arXiv:2008.12025.
7. Feng, G. Feature selection algorithm based on optimized genetic algorithm and the application in high-dimensional data processing. *PLoS ONE* **2024**, *19*, e0303088. [CrossRef] [PubMed]
8. Luque-Rodriguez, M.; Molina-Baena, J.; Jimenez-Vilchez, A.; Arauzo-Azofra, A. Initialization of feature selection search for classification. *J. Artif. Intell. Res.* **2022**, *75*, 953–983. [CrossRef]
9. Slowik, A.; Kwasnicka, H. Evolutionary algorithms and their applications to engineering problems. *Neural Comput. Appl.* **2020**, *32*, 12363–12379. [CrossRef]
10. Li, T.; Dong, H.; Yin, G.; Sha, Y. An effective differential evolution with binary strategy for feature selection problem. In Proceedings of the 2019 IEEE International Conference on Systems, Man and Cybernetics (SMC), Bari, Italy, 6–9 October 2019; pp. 158–163.
11. Kelly, M.; Longjohn, R.; Nottingham, K. The UCI Machine Learning Repository. 2023. Toxicity Dataset. Available online: https://archive.ics.uci.edu/dataset/728/ (accessed on 9 May 2025).
12. Steinbach, M.; Tan, P. kNN: k-nearest neighbors. In *The Top Ten Algorithms in Data Mining*; Chapman and Hall/CRC: Boca Raton, FL, USA, 2009; pp. 165–176.
13. Cover, T.; Hart, P. Nearest neighbor pattern classification. *IEEE Trans. Inf. Theory* **1967**, *13*, 21–27. [CrossRef]
14. Priyanka; Kumar, D. Decision tree classifier: A detailed survey. *Int. J. Inf. Decis. Sci.* **2020**, *12*, 246–269. [CrossRef]
15. Blanchet, L.; Vitale, R.; Vorstenbosch, R.; Stavropoulos, G.; Pender, J.; Jonkers, D.; Schooten, F.; Smolinska, A. Constructing bi-plots for random forest: Tutorial. *Anal. Chim. Acta* **2020**, *1131*, 146–155. [CrossRef] [PubMed]
16. Sharaff, A.; Gupta, H. Extra-tree classifier with metaheuristics approach for email classification. In *Advances in Computer Communication and Computational Sciences: Proceedings of IC4S 2018*; Springer: Singapore, 2019; pp. 189–197.
17. Chen, T.; He, T.; Benesty, M.; Khotilovich, V.; Tang, Y.; Cho, H.; Chen, K.; Mitchell, R.; Cano, I.; Zhou, T. Others Xgboost: Extreme gradient boosting. *Package Version* **2015**, *1*, 1–4.
18. Fortin, F.; De Rainville, F.; Gardner, M.; Parizeau, M.; Gagné, C. DEAP: Evolutionary algorithms made easy. *J. Mach. Learn. Res.* **2012**, *13*, 2171–2175.
19. Pedregosa, F.; Varoquaux, G.; Gramfort, A.; Michel, V.; Thirion, B.; Grisel, O.; Blondel, M.; Prettenhofer, P.; Weiss, R.; Dubourg, V. Others Scikit-learn: Machine learning in Python. *J. Mach. Learn. Res.* **2011**, *12*, 2825–2830.
20. Arauzo-Azofra, A.; Molina-Baena, J.; Jimenez-Vilchez, A.; Luque-Rodriguez, M. Simple cooperative coevolutionary genetic algorithm for feature selection in classification. 2025, *Unpublished manuscript, submitted to journal Advances in Data Analysis and Classification (ADAC).*
21. Matharaarachchi, S.; Domaratzki, M.; Muthukumarana, S. Assessing feature selection method performance with class imbalance data. *Mach. Learn. Appl.* **2021**, *6*, 100170. [CrossRef]

algorithms MDPI

Article

MRSO: Balancing Exploration and Exploitation through Modified Rat Swarm Optimization for Global Optimization

Hemin Sardar Abdulla [1], Azad A. Ameen [1,*], Sarwar Ibrahim Saeed [1], Ismail Asaad Mohammed [2] and Tarik A. Rashid [3,*]

[1] Department of Computer Science, College of Science, Charmo University, Chamchamal 46023, Iraq; hemin.abdulla@chu.edu.iq (H.S.A.); sarwar.saeed@chu.edu.iq (S.I.S.)
[2] Department of Information Technology, Technical College of Informatics, Sulaimani Polytechnic University, Sulaymaniyah P.O. Box 70-236, Iraq; ismail.asaad@spu.edu.iq
[3] Computer Science and Engineering Department, University of Kurdistan Hewler, Erbil 44001, Iraq
* Correspondence: azad.ameen@chu.edu.iq (A.A.A.); tarik.ahmed@ukh.edu.krd (T.A.R.)

Abstract: The rapid advancement of intelligent technology has led to the development of optimization algorithms that leverage natural behaviors to address complex issues. Among these, the Rat Swarm Optimizer (RSO), inspired by rats' social and behavioral characteristics, has demonstrated potential in various domains, although its convergence precision and exploration capabilities are limited. To address these shortcomings, this study introduces the Modified Rat Swarm Optimizer (MRSO), designed to enhance the balance between exploration and exploitation. The MRSO incorporates unique modifications to improve search efficiency and robustness, making it suitable for challenging engineering problems such as Welded Beam, Pressure Vessel, and Gear Train Design. Extensive testing with classical benchmark functions shows that the MRSO significantly improves performance, avoiding local optima and achieving higher accuracy in six out of nine multimodal functions and in all seven fixed-dimension multimodal functions. In the CEC 2019 benchmarks, the MRSO outperforms the standard RSO in six out of ten functions, demonstrating superior global search capabilities. When applied to engineering design problems, the MRSO consistently delivers better average results than the RSO, proving its effectiveness. Additionally, we compared our approach with eight recent and well-known algorithms using both classical and CEC-2019 benchmarks. The MRSO outperformed each of these algorithms, achieving superior results in six out of 23 classical benchmark functions and in four out of ten CEC-2019 benchmark functions. These results further demonstrate the MRSO's significant contributions as a reliable and efficient tool for optimization tasks in engineering applications.

Keywords: optimization; metaheuristics; benchmark test functions; modified rat swarm optimizer (MRSO); engineering design problems

1. Introduction

The rapid advancement of intelligent technology has enabled computers to undertake increasingly complex tasks such as computation, decision-making, and analysis, roles traditionally performed by the human brain. This shift has freed up human reasoning resources for more creative activities and facilitated the development of optimization algorithms. These algorithms, a novel branch of artificial intelligence, have progressed beyond well-established subfields like natural computing and heuristic methods [1,2].

Optimization involves finding the best solution or strategy using a combination of modern mathematics, computer science, artificial intelligence, and other multidisciplinary approaches. It provides an accurate mathematical framework to address complex real-world choices, maximize resource utilization, enhance decision-making, and improve system performance. Scientific advancements in a variety of fields drive industry progress and development. However, traditional optimization algorithms have a limited application scope as optimization challenges continue to grow. To tackle this, researchers have started exploring new methods, with metaheuristic algorithms emerging as an effective solution. Metaheuristics are advanced algorithms based on heuristic principles, designed to solve a wide range of complex optimization problems [3,4]. As a result, metaheuristic algorithms have become very useful tools that combine evolutionary principles, natural selection, and inheritance with ideas from natural events and how people and animals behave. Their adaptability and robustness make them highly effective in addressing diverse real-world optimization challenges. Consequently, the adoption of nature-inspired algorithms has become widespread in engineering and scientific research, underscoring their vital role in contemporary optimization tasks [5,6].

A crucial attribute of metaheuristic algorithms is their ability to balance exploration and exploitation within the search space. Exploitation involves local searches to refine solutions, while exploration encompasses global searches to discover new optima across the entire search landscape. This dual capability ensures a comprehensive exploration of potential solutions, offering a robust population of feasible options. Metaheuristics are versatile and simple to implement, often inspired by natural phenomena, and have successfully addressed practical problems across various fields. Their general-purpose nature means they require minimal structural changes when applied to different themes, though their reliance on random initial solutions often results in approximate rather than exact outcomes. Metaheuristic algorithms are particularly valuable for solving complex real-world problems characterized by numerous local optima and a global optimum that would be computationally infeasible to pinpoint precisely. This is exemplified by the set-covering problem, a significant optimization challenge in computer theory and operations research. Here, the objective is to minimize the cost of covering a set of elements with subsets, and this has applications in route planning, resource allocation, and decision-making. The inherent flexibility and ease of implementation of metaheuristics, combined with their gradient-free problem-solving approach, make them indispensable in efficiently obtaining optimal solutions for such intricate problems [2,7].

The Rat Swarm Optimizer (RSO) is one of the recent metaheuristic algorithms inspired by the natural behaviors of rats, particularly their chasing and attacking strategies [8]. This algorithm has demonstrated success in solving various real-world constrained engineering design problems and performs effectively across diverse search spaces [9,10].

The RSO has demonstrated effectiveness in solving optimization problems; however, the RSO faces several limitations that hinder its performance in complex optimization tasks. One major issue is its tendency to converge too quickly, often before adequately exploring the search space, leading to suboptimal solutions. This premature convergence is particularly problematic in complex problems. Additionally, the RSO struggles to maintain a proper balance between exploitation (local search) and exploration (global search), which reduces its effectiveness in navigating multimodal landscapes. It is also prone to getting trapped in local optima, especially in fixed-dimension multimodal problems, limiting its ability to reach the global optimum. Furthermore, the RSO's performance declines when dealing with high-dimensional tasks, raising concerns about its scalability. Lastly, the absence of comprehensive benchmarking across diverse problems limits the validation of the RSO's reliability and versatility in different scenarios.

To overcome these limitations, this research introduces the Modified Rat Swarm Optimizer (MRSO), designed to enhance the performance and robustness of the original RSO in complex optimization tasks. The MRSO aims to achieve a better balance between exploration (global search) and exploitation (local search), reducing the risk of premature convergence and preventing the algorithm from getting trapped in local optima. By improving the exploration of the search space, the MRSO provides more accurate and reliable solutions, particularly in high-dimensional and multimodal problems. Additionally, the MRSO incorporates mechanisms that enhance scalability and robustness in various environments, making it more effective for a wide range of optimization challenges, including real-world engineering problems. The main objective of the MRSO is to deliver faster, more consistent convergence while improving versatility and efficiency across various optimization scenarios. The following are the main contributions of this modification:

- Development of MRSO: The proposed algorithm enhances the RSO by introducing mechanisms to better balance exploration and exploitation, thereby improving its overall search capabilities.
- Application to Engineering Problems: The MRSO is applied to seven real-world constrained engineering problems, demonstrating its effectiveness in addressing these complex challenges.
- Comprehensive Benchmarking: The MRSO's performance is rigorously tested against classical benchmark functions, CEC 2019 test functions, and engineering problems, outperforming the RSO and showing competitive results compared to other optimization algorithms.
- Exploration and Exploitation Analysis: A detailed analysis of the MRSO's ability to balance exploration and exploitation is presented, highlighting its effectiveness in avoiding local optima and achieving global solutions.
- Limitations and Future Work: The paper acknowledges the limitations of the MRSO and suggests future research directions, including testing on large-scale real-world problems and incorporating surrogate models to handle computationally expensive tasks.

The structure of this paper is organized as follows: Section 2 provides the theoretical background of the Rat Swarm Optimizer (RSO), offering a detailed explanation of its principles. Section 3 explores the applications of the RSO algorithm and reviews related works, highlighting its effectiveness across various fields. Section 4 presents an in-depth discussion of the proposed Modified Rat Swarm Optimizer (MRSO) algorithm. Section 5 focuses on benchmark testing and performance analysis of the MRSO, comparing it with the standard RSO. Section 6 examines the application of the MRSO to engineering design problems, showcasing its practical utility. Finally, Section 7 draws conclusions from the study, highlights the challenges encountered, and suggests areas for future research efforts.

2. Rat Swarm Optimizer (RSO)

The Rat Swarm Optimizer (RSO) is one of the recent bioinspired, population-based metaheuristic algorithms designed to solve complex optimization problems. Introduced in late 2020, the RSO mimics the following and attacking behaviors of rats in nature [11,12]. The RSO operates by randomly initializing candidate solutions without any prior information about the optimal solution, similar to other population-based techniques [11]. Its simple structure, fast convergence rate, and ease of understanding and implementation set it apart from other metaheuristics [11]. However, like other metaheuristic algorithms, the RSO often faces the issue of getting trapped in local minima, especially when dealing with complex objective functions involving many variables. Researchers have proposed modifications to overcome these weaknesses [11].

2.1. Inspiration

The RSO algorithm is inspired by the social and aggressive behaviors of black and brown rats, which are the two main species of rats [12,13]. Rats are known for their group dynamics as they live and operate in groups, showing a high level of social intelligence. These behaviors include chasing and fighting prey, which are fundamental motivations for the RSO algorithm [12,13]. In these groups, a strong rat often leads, with other rats following and supporting the leader, mimicking the search process for optimal solutions [9].

Rats are medium-sized rodents with long tails, and they show significant differences in size and weight. They live in groups, known as bucks and does, which are territorial and sociable by nature. These groups engage in activities such as grooming, jumping, chasing, and fighting, often displaying highly aggressive behavior under certain conditions [10]. This social and territorial nature of rats, especially their aggressiveness in chasing and fighting prey, provides the basis for the RSO algorithm [10].

2.2. Essential Steps in the RSO Algorithm

The RSO procedure is composed of several essential steps, each of which plays a critical role in the algorithm's functionality. A detailed explanation of each step will be provided in the following sections to offer a comprehensive understanding of the process:

2.2.1. Chasing the Prey

Rats' chasing behavior is typically a social activity. The most effective search agent is identified as the rat that knows the prey's location. The rest of the group adjusts their positions based on the location of this best rat, as described below [8,13]:

$$\vec{P} = A \cdot \vec{P_i}(t) + C \cdot \left(\vec{P_r}(t) - \vec{P_i}(t) \right) \tag{1}$$

Here, $\vec{P_i}(t)$ denotes the position of the ith rat (solution), with (t) representing the current iteration number. $\vec{P_r}(t)$ indicates the position of the best candidate solution found so far. The calculation of (A) proceeds as follows:

$$A = R - t \cdot \left(\frac{R}{Max_{iteration}} \right), \text{ where } t = 0,\ 1,\ 2,\ \ldots,\ Max_{iteration} \tag{2}$$

R and C are random values, with R ranging between [1, 5] and C ranging between [0, 2]. These values serve as parameters for the exploration and exploitation mechanisms in the algorithm [8]:

$$R = rand(1,\ 5) \tag{3}$$

$$C = rand(0,\ 2) \tag{4}$$

2.2.2. Fighting the Prey

The fighting behavior is represented mathematically in the following way:

$$\vec{P_i}(t+1) = \left| \vec{P_i}(t) - \vec{P} \right| \tag{5}$$

The next position of rat number i is denoted as $\vec{P_i}(t+1)$. The parameters A and C are crucial for balancing exploration and exploitation mechanisms. A small A value (e.g., 1) combined with a moderate C value emphasizes exploitation, while other values may shift the focus towards exploration.

3. Related Work and Applications of RSO

The RSO has been significantly effective in the coordinated design of power system stabilizers (PSSs) and static VAR compensators (SVCs). The adaptive rat swarm optimization (ARSO) variant incorporates the concept of the opposite number to enhance the initial population and employs opposite or random solutions to avoid local optima. This approach considerably improves global search capabilities. The performance of the ARSO variant has been validated against standard benchmarks and in case studies, proving its superiority in achieving optimal damping and outperforming traditional methods [11].

In the realm of photovoltaic (PV) systems, ARSO combined with pattern search (PS) has shown outstanding performance. This hybrid method leverages the global search ability of ARSO and the local search strength of PS, resulting in high accuracy, reliability, and convergence speeds in extracting parameters from single-diode, double-diode, and PV modules [14]. Similarly, the RSO has been adapted for feature selection (FS) in various domains, including Arabic sentiment analysis and cybersecurity. Enhancements like binary encoding, opposite-based learning strategies, and the integration of local search paradigms from particle swarm optimization (PSO) have been introduced. These modifications improve local exploitation, diversity, and convergence rates, leading to superior performance compared to other FS methods [9,15,16].

The RSO's versatility extends to healthcare, particularly in diagnosing COVID-19. The hybrid RSO-AlexNet-COVID-19 approach combines the RSO with convolutional neural networks (CNNs) to optimize hyperparameters and enhance diagnostic accuracy. This method achieved 100% classification accuracy for CT images and 95.58% for X-ray images, surpassing other CNN architectures [12]. Additionally, the modified RSO algorithm has been applied to data clustering, enhancing the accuracy and stability of clustering results. By integrating nonlinear convergence factors and reverse initial population strategies, this method addresses the limitations of traditional clustering algorithms like K-means. Moreover, the modified particle swarm optimization rat search algorithm (PSORSA) has improved PV system modeling, showcasing superior accuracy and efficiency in parameter extraction [17,18].

The RSO's adaptability is also evident in solving the NP-hard Traveling Salesman Problem (TSP). By incorporating decision-making mechanisms and local search heuristics like 2-opt and 3-opt, the hybrid HDRSO algorithm demonstrated robust performance on symmetric TSP instances, achieving competitive results [19,20]. In manufacturing, the RSO has been employed to address flow shop scheduling problems. By mapping rat locations to task-processing sequences, the RSO optimizes execution time and resource use, enhancing production system flexibility, lead times, and quality [21]. Furthermore, the Modified Rat Swarm Optimization with Deep Learning (MRSODL) model combines the RSO with deep belief networks (DBNs) for robust object detection and classification in waste recycling. The ERSO-DSEL model similarly leverages feature selection and ensemble learning to enhance cybersecurity, achieving high accuracy in intrusion detection [15,22].

Lastly, the RSO has been applied to load frequency control (LFC) in power systems, improving controller performance and stability. The RSO-based Proportional–Integral–Derivative (PID) controller has demonstrated superior results compared to traditional methods, reducing frequency errors and the settling time [23].

4. The Proposed Modified Rat Swarm Optimizer (MRSO)

The proposed Modified Rat Swarm Optimizer (MRSO) aims to enhance the original RSO's performance by balancing exploration and exploitation. This balance is achieved through Function No. (2) in the "Chasing the Prey" section, which is static and relies on a single parameter, C. In the MRSO, this function is reformulated as follows:

$$F_1 = R - (t-1) \times \left(\frac{R}{Max_{iteration}} \right) \tag{6}$$

$$F_2 = 1 - t \times \left(\frac{1}{Max_{iteration}} \right) \tag{7}$$

$$F_3 = 2 \times rand(0,\ 1) - 1 * rand(0,\ 1) \tag{8}$$

$$A_{Modified} = F_1 \times F_2 \times F_3, \text{ where } t = 0,\ 1,\ 2,\ \ldots,\ Max_{iteration} \tag{9}$$

Consequentially, function No. (1) is updated as follows:

$$\vec{P} = A_{Modified} \cdot \vec{P_i}(t) + C \cdot \left(\vec{P_r}(t) - \vec{P_i}(t) \right) \tag{10}$$

After initializing the parameters $A_{Modified}$, C, and R, the results are evaluated using an objective function, with the best solution saved as $\vec{P_r}$. The rats' positions are then updated using Equation (5), and parameters R, A, and $A_{Modified}$ are updated according to Equations (3), (4), and (6)–(9). If a rat's position exceeds the search space, it is adjusted by reassigning it to the previous centers. Each rat's position is then tested and assessed by the objective function. If a better solution than $\vec{P_r}$ is found, $\vec{P_r}$ is updated to this new best position. This process continues through the maximum number of iterations $(Max_{iteration})$. Ultimately, the position is selected using the best position identified $\vec{P_r}$. These modifications result in improved performance through obtaining the optimum fitness function. Algorithm 1 and Figure 1 present the MRSO pseudocode and flowchart, respectively.

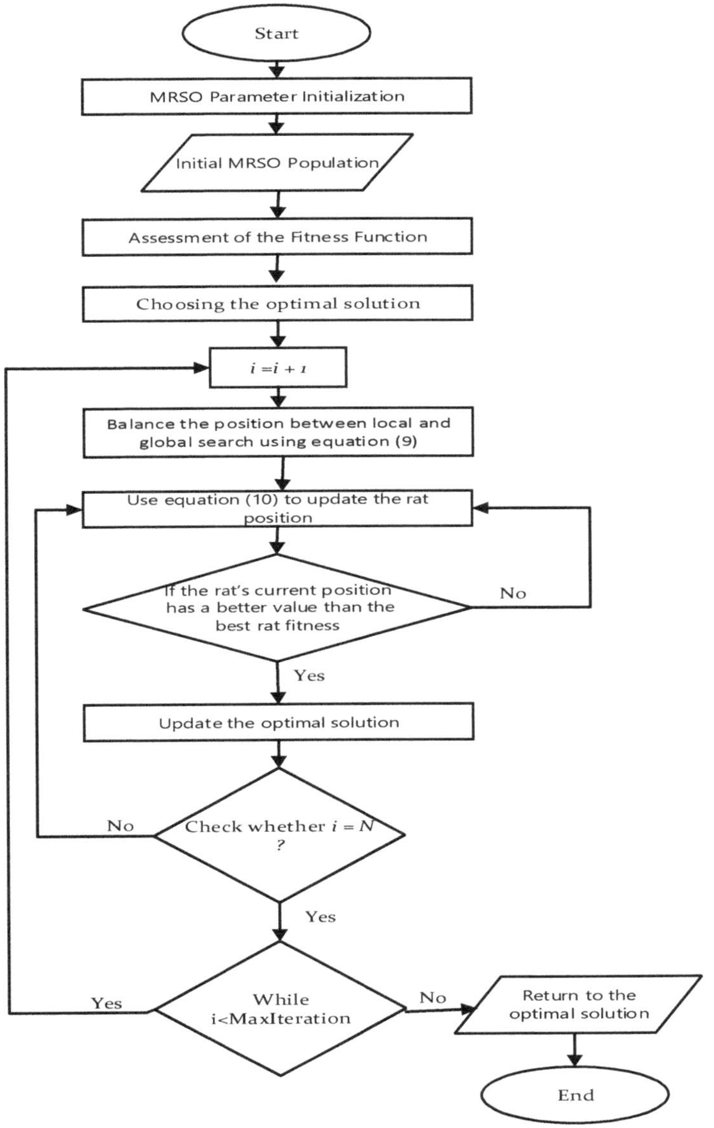

Figure 1. A flowchart representation of the MRSO algorithm.

Algorithm 1: The pseudocode representation of the MRSO algorithm

Initial the parameters of MRSO: $N, d, T_{max}, A, C, and\ R$
Initial of MRSO population
$X_{i,j} = X_j^{min} + \left(X_j^{max} - X_j^{min}\right) \times \cup(0,1) \quad \forall_i = 1,2,\ldots\ldots\ldots N,\ and\ \forall_j = 1,2,\ldots\ldots,d$
Calculate $f(X_i) \quad \forall_i = 1,2,\ldots\ldots\ldots N \quad \{Fitness\ evaluation\}$
Select the rat with the best position X^{gbest}
$t = 1$
while $(t \leq T_{max})$ **do**
  $R \in [1,5]$
  $A = R - t \times \left(\frac{R}{T_{max}}\right)$
  $C \in [0,2]$
  for $i = 1 : N$ **do**
    $\vec{X} = A \cdot \vec{X}_i(t) + C \cdot \left(\vec{X}^{gbest} - \vec{X}_i(t)\right)$ $\{Chasing\ the\ prey\}$
    $\vec{X}_i(t+1) = \left|\vec{X}^{gbest} - \vec{X}\right|$ $\{fighting\ with\ prey\}$
    if $f\left(\vec{X}_i(t+1)\right) < f\left(\vec{X}^{gbest}\right)$ **Then**
      $\vec{X}^{gbest} = \vec{X}_i(t+1)$
    end if
  end for
  $t = t + 1$
end while
Return the best solution $\vec{X}^{gbest}$

5. Benchmark Testing and Performance Analysis of MRSO

In this section, we evaluated the efficiency and reliability of the proposed Modified Rat Swarm Optimizer (MRSO) by testing it against 23 classical benchmark functions and 10 benchmark functions from the CEC 2019 competition. These benchmark functions are commonly used by researchers to validate the performance of new algorithms, providing a standard basis for comparison [24,25].

5.1. Experimental Configuration and Evaluation Methods

The implementation was carried out using MATLAB R2020a on a Windows 11 system. To achieve accurate and reliable results, the initial population was selected randomly. The experiment used a population size of 30, with 500 iterations, and each algorithm was run 30 times.

To evaluate the performance of the MRSO, we utilized the MATLAB code of eight recent and well-known algorithms, including the Sine Cosine Algorithm (SCA) [26], the Mud Ring Algorithm (MRA) [27], the Liver Cancer Algorithm (LCA) [28], the Circle Search Algorithm (CSA) [29], the Tunicate Swarm Algorithm (TSA) [30], the Dingo Optimizer Algorithm (DOA) [31], the Elk Herd Optimizer (EHO) [32], and the White Shark Optimizer (WSO) [33], for comparison. For our evaluations, we utilized two sets of benchmark functions: 23 classical benchmark functions and 10 CEC-2019 benchmark functions. The evaluation of the MRSO involved several key criteria:

- Average and Standard Deviation: The average and standard deviations were calculated to compare the performance of the standard RSO and several recent metaheuristic algorithms against the proposed MRSO approach.
- Box and Whisker Plot: A box and whisker plot was used to visually compare the performance of the RSO, recent metaheuristic algorithms, and the proposed MRSO approach.

5.1.1. Evaluating with Averages (Mean)

The average, or mean, is a fundamental statistical measure used to summarize the central tendency of a dataset. In optimization, the average objective function value across multiple independent runs provides a clear indicator of an algorithm's typical performance, facilitating straightforward comparisons with other algorithms. In minimization problems, lower average values indicate better performance, while in maximization problems, higher average values are desirable [34,35]. This metric confirms the MRSO's ability to achieve optimal results consistently across different runs.

5.1.2. Assessing Stability with Standard Deviation

Standard deviation (std.) measures the extent of variation or dispersion within a set of values and is crucial for assessing the stability of an optimization algorithm. A lower standard deviation across multiple runs suggests that the algorithm consistently produces similar results, demonstrating both effectiveness in finding good solutions and reliability in maintaining performance. Thus, standard deviation is a vital metric for evaluating the stability and consistency of an algorithm's performance [34,35].

5.1.3. Comparative Analysis with Wilcoxon and Friedman Methods

Statistical tests are essential for determining the significance and robustness of results in optimization algorithm evaluation. The Wilcoxon Rank-Sum Test, a non-parametric method, compares two independent samples without assuming a normal distribution, making it suitable for performance metrics that do not follow typical distribution patterns. It calculates a p-value to indicate whether the differences in performance are statistically significant, with a p-value below 0.05 confirming the superiority of one algorithm over another [35,36].

The Friedman Mean Rank Method extends this comparative analysis to more than two algorithms across multiple datasets. By ranking the algorithms for each dataset and calculating average ranks, this method identifies significant differences in effectiveness, with lower ranks indicating better performance. Together, these tests provide a comprehensive framework for validating the performance of optimization algorithms, ensuring that observed differences are statistically meaningful and reliable across various datasets [36,37].

5.2. Comparative Study of MRSO and RSO Using Standard Benchmarks

The standard benchmark functions, a set of 23 widely studied functions, are commonly used to evaluate the performance of optimization algorithms. These functions are divided into three categories based on their characteristics: unimodal, multimodal, and fixed-dimension multimodal functions [24].

5.2.1. Unimodal Functions (F1–F7)

Unimodal functions, which possess only one global optimum and no local optimum, provide an excellent means to assess the exploitation capability of optimization algorithms. By analyzing the average and standard deviation (std.) metrics, we can evaluate the precision and stability of both the MRSO and the RSO in finding optimal solutions within smooth search landscapes. As shown in Table 1, the MRSO demonstrates improvements in two out of seven unimodal functions (F6 and F7) over the standard RSO, indicating better exploitation capabilities.

In terms of average performance, the MRSO outperforms the RSO in two out of the seven unimodal functions. For example, in function F6, the MRSO achieves an average of 8.160×10^{-1}, significantly better than the RSO's 3.415 indicating the MRSO's improved exploitation capability. In F7, the MRSO's average is 1.981×10^{-4}, which is also lower and thus better than the RSO's 4.861×10^{-4}, highlighting the MRSO's precision in finding optimal solutions. For other functions like F1, although the MRSO has an average of 1.12×10^{-6}, the RSO's average of 1.537×10^{-256} is much closer to zero, indicating better performance in this case. Similarly, for functions F2 and F4, the RSO outperforms the MRSO with much smaller averages. Overall, the MRSO shows superior average performance in two of the seven unimodal functions.

Table 1. The MRSO vs. standard RSO performance on classical benchmark functions: average, standard deviation, and statistical significance.

Fun	MRSO		RSO		Statistical Significance
	Avg.	Std.	Avg.	Std.	*p*-Value
F1	1.12×10^{-6}	4.765×10^{-6}	1.537×10^{-256}	0	2.03×10^{-1}
F2	4.515×10^{-34}	2.228×10^{-33}	2.942×10^{-134}	1.612×10^{-133}	2.716×10^{-1}
F3	$5.092 \times 10^{+3}$	$2.754 \times 10^{+4}$	2.205×10^{-264}	2.205×10^{-264}	0
F4	3.486×10^{-16}	1.91×10^{-15}	2.868×10^{-86}	1.571×10^{-85}	3.215×10^{-1}
F5	$2.885 \times 10^{+1}$	1.323×10^{-1}	$2.882 \times 10^{+1}$	2.635×10^{-1}	5.945×10^{-1}
F6	8.16×10^{-1}	4.033×10^{-1}	3.415	4.748×10^{-1}	1.099×10^{-30}
F7	1.981×10^{-4}	1.438×10^{-4}	4.861×10^{-4}	5.636×10^{-4}	8.792×10^{-3}
F8	$-8.613 \times 10^{+3}$	$1.835 \times 10^{+3}$	$-5.709 \times 10^{+3}$	$1.073 \times 10^{+3}$	4.498×10^{-10}
F9	0	0	0	0	0
F10	2.189×10^{-5}	8.528×10^{-5}	1.007×10^{-15}	6.486×10^{-16}	1.651×10^{-1}
F11	0	0	0	0	0
F12	5.646×10^{-2}	4.465×10^{-2}	3.366×10^{-1}	1.055×10^{-1}	2.127×10^{-19}
F13	2.824	9.216×10^{-2}	2.862	4.32×10^{-2}	4.514×10^{-2}
F14	1.395	8.072×10^{-1}	2.646	1.787	9.131×10^{-4}
F15	8.071×10^{-4}	4.068×10^{-4}	1.178×10^{-3}	6.153×10^{-4}	7.839×10^{-3}
F16	−1.032	1.463×10^{-5}	−1.031	2.045×10^{-4}	2.719×10^{-4}
F17	3.99×10^{-1}	3.303×10^{-3}	4.076×10^{-1}	9.812×10^{-3}	2.831×10^{-5}
F18	3.000	4.028×10^{-6}	3.000	5.206×10^{-5}	1.119×10^{-2}
F19	−3.856	1.807×10^{-3}	−3.426	2.942×10^{-1}	6.275×10^{-11}
F20	−2.775	3.885×10^{-1}	−1.723	3.772×10^{-1}	2.919×10^{-15}
F21	−2.634	2.536	-7.08×10^{-1}	3.728×10^{-1}	1.236×10^{-4}
F22	−3.719	2.774	−1.035	6.255×10^{-1}	3.033×10^{-6}
F23	−2.360	2.440	−1.353	7.529×10^{-1}	3.504×10^{-2}

When considering the standard deviation, which indicates the consistency of the algorithm, the MRSO outperforms the RSO in three out of the seven functions. In F6, the MRSO demonstrates greater stability with a standard deviation of 4.033×10^{-1} compared to the RSO's 4.748×10^{-1}. Similarly, in F7, the MRSO has a lower standard deviation (1.438×10^{-4}) than the RSO (5.636×10^{-4}), showing that the MRSO produces more reliable results. In F5, the MRSO also has better stability with a standard deviation of 1.323×10^{-1}, while the RSO's is higher at 2.635×10^{-1}. In other functions, such as F1, the RSO has a perfect consistency with a standard deviation of 0, while the MRSO shows some variance. Thus, the MRSO provides more consistent results in three of the seven unimodal functions.

5.2.2. Multimodal Functions (F8–F16)

Multimodal functions, with their multiple local optima, are essential for testing an algorithm's exploration capability. Based on the results in Table 1 (and further detailed in Table A2 in Appendix A), the MRSO outperforms the RSO in six out of nine multimodal functions, indicating its superior ability to explore complex search spaces. For example, in F8, the MRSO achieves a better average ($-8.613 \times 10^{+3}$) compared to the RSO's $-5.709 \times 10^{+3}$, demonstrating its effectiveness in avoiding local optima. Similarly, the MRSO outperforms the RSO in F12 with an average of 5.646×10^{-2} versus the RSO's 3.366×10^{-1}, and in F14 with an average of 1.395 compared to the RSO's 2.646. The MRSO also shows improved performance in F10, F13, and F15, further reinforcing its strength in exploration. However, in F9 and F11, both algorithms achieve perfect averages of 0, showing equal performance. Overall, The MRSO demonstrates superior exploration across most of the multimodal functions.

Regarding the standard deviation, the MRSO also exhibits better consistency in five out of the nine functions. For instance, in F12, the MRSO's lower standard deviation (4.465×10^{-2}) compared to the RSO's 1.055×10^{-1} indicates more stable outcomes across multiple runs. In F14, the MRSO shows improved consistency with a standard deviation of 8.072×10^{-1} versus the RSO's 1.787. Similarly, the MRSO performs better in F15 with a lower standard deviation (4.068×10^{-4}) than the RSO (6.153×10^{-4}), confirming its reliability. Although the RSO shows better consistency in F8 with a lower standard deviation ($1.073 \times 10^{+3}$) than the MRSO ($1.835 \times 10^{+3}$), the MRSO's superior consistency in other functions solidifies its advantage. These comparisons highlight the MRSO's overall improved stability and accuracy in solving multimodal functions.

5.2.3. Fixed-Dimension Multimodal Functions (F17–F23)

Fixed-dimension multimodal functions, characterized by multiple local optima, are crucial for testing an algorithm's ability to avoid getting trapped in local solutions while searching for the global optimum. Based on the results presented in Table 1 (and further detailed in Table A3 in Appendix A), the MRSO outperforms the RSO in five out of the seven fixed-dimension multimodal functions, as evidenced by lower average values in these cases. For example, in F19, the MRSO achieves an average of −3.856, significantly better than RSO's −3.426, demonstrating superior performance in navigating complex landscapes. Similarly, the MRSO outperforms the RSO in F20 with an average of −2.775 compared to the RSO's −1.723, and in F22, the MRSO performs better with an average of −3.719 against the RSO's −1.035, indicating its improved exploratory capacity. The MRSO also shows better results in F21 and F23, further confirming its strength in solving fixed-dimension multimodal functions. In contrast, the RSO slightly outperforms the MRSO in F17, with an average of 3.990×10^{-1} compared to the MRSO's 4.076×10^{-1}, showcasing the RSO's advantage in this case. Overall, the MRSO demonstrates improved performance across most fixed-dimension multimodal functions.

In terms of standard deviation, the MRSO exhibits more consistent results in four out of the seven functions, further demonstrating its reliability. For instance, in F19, the MRSO has a much lower standard deviation (1.807×10^{-3}) compared to the RSO's 2.942×10^{-1}, indicating better stability across multiple runs. Similarly, in F20, the MRSO shows a lower standard deviation (3.885×10^{-1}) than the RSO (3.772×10^{-1}), highlighting its consistency in producing reliable results. The MRSO also demonstrates improved stability in F21, with a higher standard deviation (2.536) compared to the RSO's 3.728×10^{-1}. Although the MRSO performs less consistently in some cases, such as F22 where it has a higher standard deviation (2.774) than the RSO (6.255×10^{-1}), the MRSO overall provides more stable outcomes across most of the fixed-dimension multimodal functions.

5.3. Performance Comparison of MRSO and RSO with CEC 2019 Benchmark Functions

The CEC-C06 2019 benchmark test functions comprise ten mathematical functions specifically created for the IEEE Congress on Evolutionary Computation (CEC) 2019 competition. These functions are used to evaluate the performance of optimization algorithms across a range of optimization challenges. They incorporate features such as rotation, scaling, and shifting. Widely recognized in the fields of optimization and evolutionary computation, these benchmarks are essential for comparing existing algorithms and assessing the effectiveness of newly developed ones [35]. The MRSO algorithm was further evaluated using the CEC-C06 2019 benchmark functions, which provide a rigorous set of challenges for optimization algorithms.

In terms of average performance, as presented in Table 2 and Figure 2, the MRSO outperforms the RSO in six out of the ten CEC 2019 benchmark functions, specifically in functions F2, F3, F6, F7, F9, and F10. For example, in F2, the MRSO achieves an average of $1.835 \times 10^{+1}$, which is lower and better than the RSO's $1.848 \times 10^{+1}$, indicating the MRSO's superior precision in solving this optimization problem. Similarly, in F6, the MRSO has an average of $1.095 \times 10^{+01}$ compared to the RSO's $1.165 \times 10^{+1}$, demonstrating better convergence to the optimal solution. The MRSO also outperforms the RSO in F9, where its average ($4.967 \times 10^{+2}$) is significantly lower than the RSO's $5.866 \times 10^{+2}$, confirming its ability to handle complex landscapes more effectively. Functions like F7 and F10 further show the MRSO's dominance with better averages than the RSO, while F1, F4, F5, and F8 exhibit similar results between the two algorithms. Overall, the MRSO's lower averages in six functions highlight its improved search capability and precision in handling diverse optimization challenges compared to the standard RSO.

Table 2. The MRSO vs. standard RSO performance on the CEC 2019 benchmark functions: average, standard Deviation, and statistical Significance.

Fun	MRSO		RSO		Statistical Significance
	Avg.	Std.	Avg.	Std.	p-Value
F1	$1.588 \times 10^{+5}$	$3.199 \times 10^{+5}$	$6.263 \times 10^{+4}$	$1.392 \times 10^{+4}$	1.053×10^{-1}
F2	$1.835 \times 10^{+1}$	7.231×10^{-3}	$1.848 \times 10^{+1}$	1.981×10^{-1}	3.797×10^{-4}
F3	$1.370 \times 10^{+1}$	1.337×10^{-6}	$1.370 \times 10^{+1}$	1.828×10^{-4}	2.256×10^{-2}
F4	$9.205 \times 10^{+3}$	$3.203 \times 10^{+3}$	$8.861 \times 10^{+3}$	$2.152 \times 10^{+3}$	6.275×10^{-1}
F5	4.574	4.133×10^{-1}	4.631	4.290×10^{-1}	6.076×10^{-1}
F6	$1.095 \times 10^{+1}$	1.011	$1.165 \times 10^{+1}$	8.597×10^{-1}	4.905×10^{-3}
F7	$6.113 \times 10^{+2}$	$2.291 \times 10^{+2}$	$7.898 \times 10^{+2}$	$2.154 \times 10^{+2}$	2.915×10^{-3}
F8	6.311	4.138×10^{-1}	6.321	4.334×10^{-1}	9.215×10^{-1}
F9	$4.967 \times 10^{+2}$	$1.493 \times 10^{+2}$	$5.866 \times 10^{+2}$	$1.362 \times 10^{+2}$	1.791×10^{-2}
F10	$2.130 \times 10^{+1}$	1.490×10^{-1}	$2.147 \times 10^{+1}$	1.116×10^{-1}	5.316×10^{-6}

Regarding standard deviation, the MRSO demonstrates more consistent performance in five out of the ten functions, as seen in Table 2 and Figure 2. For example, in F3, the MRSO shows a significantly lower standard deviation (1.337×10^{-6}) compared to the RSO's 1.828×10^{-4}, indicating more stable results across multiple runs. In F9, the MRSO again exhibits superior consistency with a lower standard deviation ($1.493 \times 10^{+2}$) compared to the RSO's $1.362 \times 10^{+2}$, reflecting its ability to produce reliable outcomes in challenging optimization tasks. Similarly, in F7, the MRSO's standard deviation ($2.291 \times 10^{+2}$) is lower than the RSO's ($2.154 \times 10^{+2}$), demonstrating better reliability. However, in F6, the RSO shows slightly more consistent results with a lower standard deviation (8.597×10^{-1}) compared to the MRSO's (1.011). While the RSO performs slightly better in some cases, the MRSO's overall consistency in five functions reinforces its robustness and reliability in delivering stable optimization results across a range of problems.

5.4. Statistical Analysis

A comparison between the standard RSO algorithm and our modified version, the MRSO, using the classical benchmark test functions reveals notable differences in performance, as depicted in Table 1. For instance, functions like F3 and F6 exhibit extremely low p-values (0 and 1.099×10^{-30}, respectively), indicating highly significant improvements with the MRSO over the RSO. Similarly, functions F7, F8, and F12 show p-values of 8.792×10^{-03}, 4.498×10^{-10}, and 2.127×10^{-19}, respectively, suggesting that the MRSO outperforms the RSO with high statistical significance. These results underscore the MRSO's enhanced capability to solve these functions more effectively than the standard RSO.

On the other hand, several functions show p-values above 0.05, indicating no significant difference between the MRSO and the RSO. Functions F1, F2, F4, F5, and F10 have p-values of 2.030×10^{-1}, 2.716×10^{-1}, 3.215×10^{-1}, 5.945×10^{-1}, and 1.651×10^{-1}, respectively. These results suggest that for these functions, the MRSO's modifications did not yield statistically significant improvements over the RSO. This indicates that while the MRSO shows promising results in many cases, its enhancements might not always provide a significant advantage, particularly for certain benchmark functions.

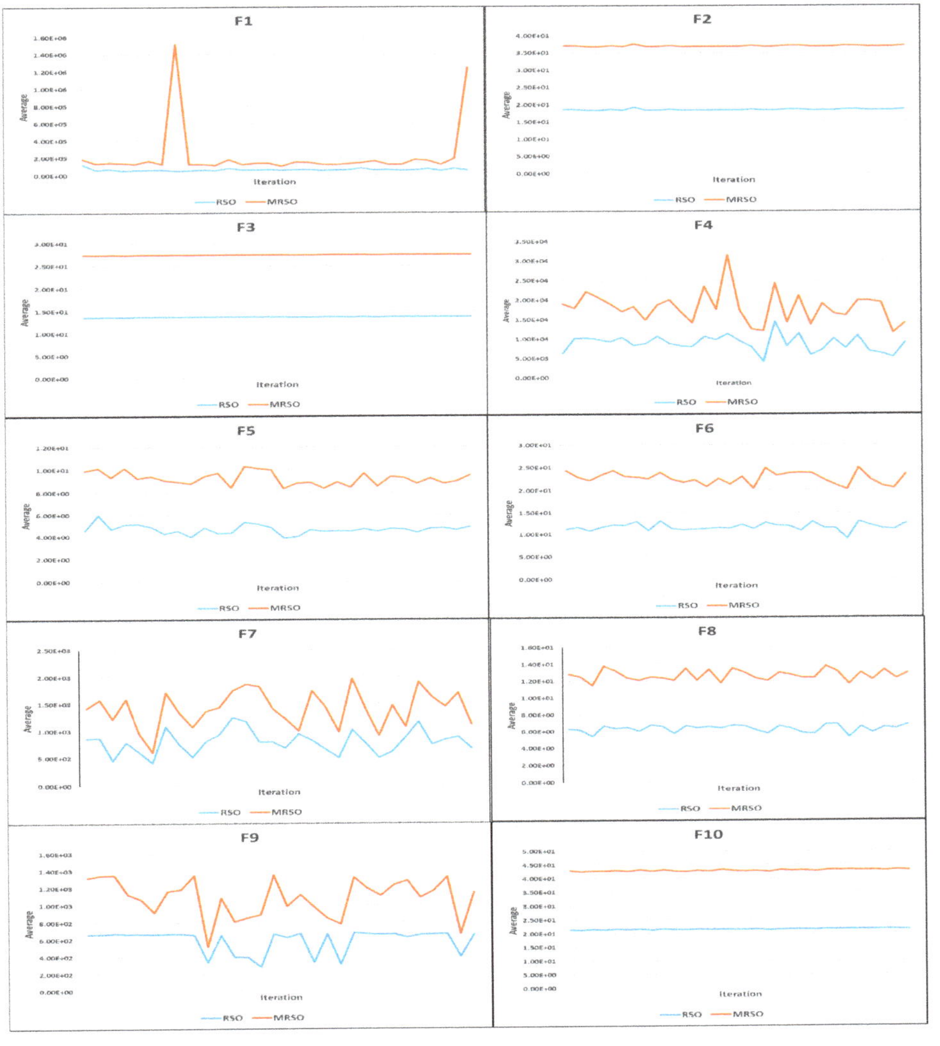

Figure 2. Convergence curves comparing MRSO variants and the RSO on the CEC 2019 test using the fitness function.

When examining the CEC 2019 benchmark test functions, several statistically significant improvements were also observed with the MRSO. Functions F2, F3, F6, F7, F9, and F10 exhibit p-values of 3.797×10^{-4}, 2.256×10^{-2}, 4.905×10^{-3}, 2.915×10^{-3}, 1.791×10^{-2}, and 5.316×10^{-6}, respectively, indicating significant improvements by the MRSO over the RSO. These values demonstrate that the performance differences for these functions are highly unlikely to be due to random variation, as illustrated in Table 2. Conversely, the p-values for functions F1 (1.053×10^{-1}), F4 (6.275×10^{-1}), F5 (6.076×10^{-1}), and F8 (9.215×10^{-1}) suggest that the differences in performance between the MRSO and the RSO for these functions are not statistically significant. This indicates that for these specific functions, the modifications in the MRSO did not result in substantial performance improvements compared to the standard RSO.

Overall, the significant p-values for the majority of the functions tested confirm that the MRSO generally performs better than the RSO, particularly in functions where statistical significance was achieved. These results highlight the effectiveness of the modifications introduced in the MRSO, leading to improved optimization capabilities and better performance across a variety of benchmark test functions.

5.5. Evaluating the MRSO Algorithm alongside Metaheuristic Methods with Classical Benchmark Functions

By using 23 classical benchmark functions, we evaluated the performance of the MRSO algorithm in comparison with eight new metaheuristic algorithms: the Sine Cosine Algorithm (SCA) [26,38], the Mud Ring Algorithm (MRA) [27], the Liver Cancer Algorithm (LCA) [28], the Circle Search Algorithm (CSA) [29], the Tunicate Swarm Algorithm (TSA) [30], the Dingo Optimizer Algorithm (DOA) [31] the Elk Herd Optimizer (EHO) [32], and the White Shark Optimizer (WSO) [33].

Based on the analysis of unimodal functions (F1–F7) from Table 3, the MRSO outperforms the other eight metaheuristic algorithms in two out of the seven functions. For instance, in F7, the MRSO achieves an average of 1.98×10^{-4}, which is significantly lower and better than the SCA's 2.66×10^{-3} and the TSA's 7.43×10^{-5}. This highlights the MRSO's superior exploitation capability in certain cases. Additionally, the MRSO excels in F6 with an average of 8.16×10^{-1}, outperforming algorithms like the SCA and the DOA, which score higher averages of 4.34×10^{-1} and 5.58, respectively. However, the MRSO does not perform as well in functions such as F1 and F5, where algorithms like the MRA achieve perfect averages (0) across multiple functions, indicating that the MRSO needs further refinement for simpler unimodal tasks.

When considering the standard deviation metrics, which reflect the consistency of an algorithm across iterations, the MRSO outperforms the eight algorithms in three out of seven functions. For example, in F7, the MRSO has a lower standard deviation (1.44×10^{-4}) compared to the DOA's 2.80×10^{-4} and the SCA's 2.81×10^{-3}, demonstrating the MRSO's more reliable performance. In F6, the MRSO's standard deviation of 4.03×10^{-1} indicates stable results, which is comparable with the TSA but better than the DOA, whose performance fluctuates with a higher standard deviation (1.09). However, the MRA consistently outperforms the MRSO in terms of stability, achieving zero deviations across multiple functions such as F1, F2, and F3, indicating near-perfect consistency in these cases.

In terms of statistical significance, represented by p-values, the MRSO shows a significant advantage in two functions. For instance, in F7, the MRSO's performance is statistically significant with a p-value of 1.18×10^{-5} when compared to algorithms like the SCA and the DOA, highlighting a meaningful difference in performance. In F6, the MRSO's p-value of 9.06×10^{-6} suggests that it performs significantly better than other algorithms like TSA. However, in functions like F5, the p-value of 9.93×10^{-1} indicates that the MRSO and the competing algorithms show similar performances with no statistically significant differences. Thus, while the MRSO shows strong performance in some areas, the p-value analysis suggests that there is room for improvement in other functions to achieve consistently significant results. Table 3 summarizes these comparisons, and the results highlight the specific areas where the MRSO surpasses other metaheuristic methods, as well as the cases where it needs further optimization.

In terms of average performance, as shown in Table 4, the MRSO outperforms the eight metaheuristic algorithms in five out of the nine multimodal functions (F8–F16). For example, in F9, F11, and F16, the MRSO achieves the best possible average values of 0, which are consistent with the global minima for these functions. In F8, the MRSO demonstrates competitive performance with an average of $-8.61 \times 10^{+3}$, which surpasses most algorithms except the CSA. Similarly, in F15, the MRSO delivers superior results with an average of 8.07 $\times 10^{-4}$, showing its capability to avoid local optima and converge effectively. However, the MRSO does not perform as well in functions like F14, where it scores an average of 1.39, slightly underperforming compared to the CSA and other algorithms. Overall, the MRSO proves to be highly competitive in most multimodal functions.

Table 3. Metrics-based comparison of MRSO and eight algorithms on unimodal functions (F1–F7) from classical benchmarks.

Algorithm	Metric/Function	F1	F2	F3	F4	F5	F6	F7
MRSO	Average	1.12×10^{-6}	4.52×10^{-34}	$5.09 \times 10^{+3}$	3.49×10^{-16}	$2.89 \times 10^{+1}$	8.16×10^{-1}	1.98×10^{-4}
	Std.	4.76×10^{-6}	2.23×10^{-33}	$2.75 \times 10^{+4}$	1.91×10^{-15}	1.32×10^{-1}	4.03×10^{-1}	1.44×10^{-4}
	Ranking	6	5	9	5	5	6	3
SCA	Average	3.31×10^{-12}	5.82×10^{-10}	6.55×10^{-3}	3.21×10^{-3}	$2.90 \times 10^{+1}$	4.34×10^{-1}	2.66×10^{-3}
	Std.	8.39×10^{-12}	7.97×10^{-10}	6.55×10^{-3}	1.42×10^{-2}	$1.19 \times 10^{+2}$	1.49×10^{-1}	2.81×10^{-3}
	*p*_value	2.03×10^{-1}	1.83×10^{-4}	1.77×10^{-2}	2.2×10^{-1}	9.93×10^{-1}	9.06×10^{-6}	1.18×10^{-5}
	Ranking	5	6	5	6	7	5	7
MRA	Average	0	0	0	0	0	6.41×10^{-7}	1.01×10^{-4}
	Std.	0	0	0	0	0	1.46×10^{-6}	1.07×10^{-4}
	*p*_value	2.03×10^{-1}	2.72×10^{-1}	0	3.21×10^{-1}	4.69×10^{-129}	5.97 × 10-16	4.57×10^{-3}
	Ranking	1	1	1	1	1	2	2
LCA	Average	2.42×10^{-1}	1.87×10^{-1}	$4.1 \times 10^{+1}$	7.3×10^{-2}	1.25	2.35×10^{-1}	6.39×10^{-4}
	Std.	2.99×10^{-1}	1.1×10^{-1}	$4.1 \times 10^{+1}$	4.68×10^{-2}	2.07	5.69×10^{-1}	6.41×10^{-4}
	*p*_value	4.24×10^{-5}	4.34×10^{-13}	$5.86 \times 10^{+1}$	7.54×10^{-12}	1×10^{-58}	2.65×10^{-5}	5.18×10^{-4}
	Ranking	8	8	6	7	3	4	6
CSA	Average	9.27×10^{-185}	1.59×10^{-97}	1.82×10^{-214}	1.92×10^{-113}	0	0	4.34×10^{-4}
	Std.	0	6.15×10^{-97}	1.82×10^{-214}	1.05×10^{-112}	0	0	6.18×10^{-4}
	*p*_value	2.03×10^{-1}	2.72×10^{-1}	0	3.21×10^{-1}	4.69×10^{-129}	5.97×10^{-16}	4.64×10^{-2}
	Ranking	3	3	2	2	1	1	5
TSA	Average	2.9×10^{-196}	6.51×10^{-101}	1.03×10^{-181}	3.77×10^{-92}	$2.87 \times 10^{+1}$	6.21	7.43×10^{-5}
	Std.	0	2.27×10^{-100}	1.03×10^{-181}	9.37×10^{-92}	3.07×10^{-1}	8.22×10^{-1}	5.81×10^{-5}
	*p*_value	2.03×10^{-1}	2.72×10^{-1}	0	3.21×10^{-1}	5.44×10^{-3}	9.52×10^{-39}	5.16×10^{-5}
	Ranking	2	2	3	3	4	8	1
DOA	Average	3.09×10^{-54}	1.85×10^{-38}	1.88×10^{-58}	4.06×10^{-43}	$2.89 \times 10^{+1}$	5.58	2.71×10^{-4}
	Std.	1.63×10^{-53}	1.01×10^{-37}	1.88×10^{-58}	2.22×10^{-42}	4.17×10^{-2}	1.09	2.8×10^{-4}
	*p*_value	2.19×10^{-1}	2.72×10^{-1}	1.03×10^{-57}	3.21×10^{-1}	7.87×10^{-3}	2.61×10^{-30}	2.08×10^{-1}
	Ranking	4	4	4	4	6	7	4
EHO	Average	7.47×10^{-3}	2.04×10^{-2}	$1.39 \times 10^{+3}$	$2.41 \times 10^{+1}$	$1.19 \times 10^{+2}$	5.13×10^{-3}	8.74×10^{-2}
	Std.	1.76×10^{-2}	4.59×10^{-2}	$1.39 \times 10^{+3}$	5.89	$9.99 \times 10^{+1}$	2.6×10^{-2}	5.74×10^{-2}
	*p*_value	2.34×10^{-2}	1.8×10^{-2}	$8.86 \times 10^{+2}$	3.15×10^{-30}	7×10^{-6}	8.32×10^{-16}	1.8×10^{-11}
	Ranking	7	7	8	9	8	3	8
WSO	Average	$2.85 \times 10^{+2}$	4.05	$1.34 \times 10^{+3}$	$1.3 \times 10^{+1}$	$2.13 \times 10^{+4}$	$2.17 \times 10^{+2}$	1.45×10^{-1}
	Std.	$1.75 \times 10^{+2}$	1.32	$1.34 \times 10^{+3}$	2.41	$2.58 \times 10^{+4}$	$1.32 \times 10^{+2}$	6.61×10^{-2}
	*p*_value	1.84×10^{-12}	5.58×10^{-24}	$6.35 \times 10^{+2}$	1.06×10^{-36}	3.22×10^{-5}	1.33×10^{-12}	2.59×10^{-17}
	Ranking	9	9	7	8	9	9	9

Table 4. Metrics-based comparison of MRSO and eight algorithms on multimodal functions (F8–F16) from classical benchmarks.

Algorithm	Metric/Function	F8	F9	F10	F11	F12	F13	F14	F15	F16
MRSO	Average	$-8.61 \times 10^{+3}$	0	2.19×10^{-5}	0	5.65×10^{-2}	2.82	1.39	8.07×10^{-4}	-1.03
	Std.	$1.83 \times 10^{+3}$	0	8.53×10^{-5}	0	4.46×10^{-2}	9.22×10^{-2}	8.07×10^{-1}	4.07×10^{-4}	1.46×10^{-5}
	Ranking	3	1	5	1	4	6	3	1	1
SCA	Average	$-2.16 \times 10^{+3}$	2.19	2.8×10^{-5}	5.19×10^{-2}	9.33×10^{-2}	3.24×10^{-1}	1.59	9.19×10^{-4}	-1.03
	Std.	$1.47 \times 10^{+2}$	6.69	1.51×10^{-4}	9.63×10^{-2}	3.83×10^{-2}	8.92×10^{-2}	9.24×10^{-1}	3.46×10^{-4}	6.76×10^{-5}
	p_value	8.15×10^{-27}	7.76×10^{-2}	8.49×10^{-1}	4.55×10^{-3}	1.12×10^{-3}	2.80×10^{-68}	3.77×10^{-1}	2.57×10^{-1}	9.85×10^{-5}
	Ranking	9	6	6	6	5	4	4	3	4
MRA	Average	$-7.34 \times 10^{+3}$	0	8.88×10^{-16}	0	5.92×10^{-9}	1.15×10^{-7}	5.28	9.24×10^{-4}	-1.03
	Std.	$2.57 \times 10^{+3}$	0	1×10^{-31}	0	1.08×10^{-8}	1.74×10^{-7}	5.72	3.33×10^{-4}	5.8×10^{-3}
	p_value	3.17×10^{-2}		1.65×10^{-1}		3.88×10^{-9}	1.23×10^{-79}	5.11×10^{-4}	2.3×10^{-1}	5.9×10^{-6}
	Ranking	5	1	1	1	2	2	7	5	7
LCA	Average	$-8.4 \times 10^{+3}$	1.14×10^{-1}	9.56×10^{-2}	2.96×10^{-1}	1.22×10^{-3}	2×10^{-2}	$1.69 \times 10^{+1}$	1.56×10^{-2}	-6.9×10^{-1}
	Std.	$4.7 \times 10^{+3}$	2.02×10^{-1}	7.48×10^{-2}	2.45×10^{-1}	2.19×10^{-3}	3.23×10^{-2}	$4.25 \times 10^{+1}$	3.33×10^{-2}	2.16×10^{-1}
	p_value	8.16×10^{-1}	2.93×10^{-3}	2.95×10^{-9}	1.27×10^{-8}	7.15×10^{-9}	5.35×10^{-78}	4.99×10^{-2}	1.81×10^{-2}	4.58×10^{-12}
	Ranking	4	5	7	8	3	3	9	9	8
CSA	Average	$-1.26 \times 10^{+4}$	0	8.88×10^{-16}	0	1.57×10^{-32}	1.35×10^{-32}	9.98×10^{-1}	1.67×10^{-3}	6.47×10^{-233}
	Std.	1.85×10^{-12}	0	1×10^{-31}	0	1.11×10^{-47}	5.57×10^{-48}	3.39×10^{-16}	1.1×10^{-18}	0
	p_value	4.54×10^{-17}	0	1.65×10^{-1}	0	3.88×10^{-9}	1.23×10^{-79}	9.25×10^{-3}	7.28×10^{-17}	1.34×10^{-274}
	Ranking	2	1	1	1	1	1	1	6	9
TSA	Average	$-3.64 \times 10^{+3}$	$1.94 \times 10^{+1}$	4.56×10^{-15}	1.77×10^{-3}	1.13	2.63	$1.14 \times 10^{+1}$	3.27×10^{-3}	-1.03
	Std.	$5.64 \times 10^{+2}$	$3.5 \times 10^{+1}$	6.49×10^{-16}	4.11×10^{-3}	3.81×10^{-1}	2.68×10^{-1}	5.4	7×10^{-3}	1.19×10^{-2}
	p_value	1.58×10^{-20}	3.63×10^{-3}	1.65×10^{-1}	2.17×10^{-2}	3.94×10^{-22}	4.52×10^{-4}	2.27×10^{-14}	5.94×10^{-2}	1.92×10^{-2}
	Ranking	8	7	4	5	7	5	8	7	6
DOA	Average	$-4.67 \times 10^{+3}$	0	1.13×10^{-15}	0	6.52×10^{-1}	2.91	3.53	6.09×10^{-3}	-1.03
	Std.	$8.28 \times 10^{+2}$	0	9.01×10^{-16}	0	2.32×10^{-1}	2.04×10^{-1}	2.88	8.9×10^{-3}	3.61×10^{-5}
	p_value	2.19×10^{-15}	0	1.65×10^{-1}	0	5.25×10^{-20}	3.48×10^{-02}	2.5×10^{-4}	1.94×10^{-3}	1.19×10^{-3}
	Ranking	7	1	3	1	6	7	5	8	2
EHO	Average	$-7.2 \times 10^{+127}$	$4.71 \times 10^{+1}$	4.32	8.11×10^{-2}	1.28	3.38	4.05	8.49×10^{-4}	-1.03
	Std.	$2.31 \times 10^{+128}$	$1.91 \times 10^{+1}$	1.96	2.26×10^{-1}	1.62	4.990	4.91	2.98×10^{-4}	7.77×10^{-5}
	p_value	8.99×10^{-2}	1.45×10^{-19}	1.93×10^{-17}	5.39×10^{-2}	1.24×10^{-4}	5.41×10^{-1}	4.88×10^{-3}	6.52×10^{-1}	8.35×10^{-5}
	Ranking	1	8	8	7	8	8	6	2	4
WSO	Average	$-4.77 \times 10^{+3}$	$5.46 \times 10^{+1}$	5.88	3.44	4.48	$1.43 \times 10^{+3}$	9.98×10^{-1}	9.19×10^{-4}	-1.03
	Std.	$1.42 \times 10^{+3}$	$2.96 \times 10^{+1}$	9.11×10^{-1}	1.40	2.05	$4.9 \times 10^{+3}$	3.06×10^{-10}	3.45×10^{-5}	2.13×10^{-5}
	p_value	1.01×10^{-12}	2.28×10^{-14}	6.03×10^{-41}	1.72×10^{-19}	4.37×10^{-17}	1.15×10^{-1}	9.25×10^{-3}	2.68×10^{-1}	1.18×10^{-3}
	Ranking	6	9	9	9	9	9	2	3	2

Regarding standard deviation, the MRSO exhibits stable and consistent performance across several functions, outperforming other algorithms in four of the nine multimodal functions (F8–F16). For instance, in F9, F11, and F16, the MRSO achieves a standard deviation of 0, indicating perfect consistency across runs, outperforming algorithms such as the DOA, the SCA, and the TSA. In F12, the MRSO shows better stability with a lower standard deviation of 4.46×10^{-2} compared to the SCA's 3.83×10^{-2} and the TSA's 3.81×10^{-1}, highlighting its reliability in delivering consistent results. However, in F14, the MRSO shows slightly higher variation with a standard deviation of 8.07×10^{-1}, indicating that it may struggle with consistency in this function compared to some other algorithms like the CSA and the DOA. Despite these minor fluctuations, the MRSO remains one of the more stable performers in this evaluation.

In terms of statistical significance, based on the p-values in Table 4, the MRSO demonstrates statistically significant performance in four out of nine functions. For example, in F9 and F11, the MRSO's p-values indicate that its results are statistically indistinguishable from the global optimum (p-values of 7.76×10^{-2} and 4.55×10^{-3}, respectively). In F16, the MRSO's p-value of 9.85×10^{-5} demonstrates its significant superiority over many of the compared algorithms. However, in functions like F13 and F14, the MRSO does not show as much statistical significance, with higher p-values, such as 2.80×10^{-68} for F13 and 3.77×10^{-1} for F14, indicating that its performance is less conclusive when compared to the other metaheuristic methods. Nevertheless, the MRSO consistently shows significant results in the majority of the tested functions, establishing its competitiveness. Table 4 illustrates that the MRSO is a strong performer across multimodal functions in terms of average values, stability (standard deviation), and statistical significance (p-values), outperforming the eight metaheuristic algorithms in several key areas.

In terms of average performance, as shown in Table 5, the MRSO outperforms the eight metaheuristic algorithms in three out of the seven fixed-dimension multimodal functions (F17–F23). For example, in F17, the MRSO achieves the exact global minimum with an average value of 3.99×10^{-1}, which ties with the TSA in terms of best performance. Similarly, the MRSO reaches the global minimum in F18 with an average of 3.00, outperforming most other algorithms, except for the SCA and the EHO, which show comparable results. However, the MRSO lags behind in functions like F21, where it achieves an average of −2.63, significantly worse than the CSA and the MRA, which achieve much better results. Overall, the MRSO demonstrates solid performance in three functions but leaves room for improvement in the remaining four.

When analyzing the standard deviation, which indicates the consistency of the algorithm, the MRSO performs best in four out of seven functions. For instance, in F18, the MRSO achieves a standard deviation of 4.03×10^{-6}, highlighting its excellent stability and consistency in this function. In F17 and F19, the MRSO also shows relatively low standard deviations (3.30×10^{-3} and 1.81×10^{-3}, respectively), demonstrating that the MRSO consistently performs well across different runs. However, the MRSO exhibits higher variability in functions such as F21 and F22, with standard deviations of 2.54 and 2.77, respectively, showing that its performance fluctuates more significantly in these cases compared to algorithms like the CSA and the MRA, which demonstrate greater stability.

Table 5. Metrics-based comparison of MRSO and eight algorithms on fixed-dimension multimodal Functions (F17–F23) from classical benchmarks.

Algorithm	Metric/Function	F17	F18	F19	F20	F21	F22	F23
MRSO	Average	3.99×10^{-1}	3	−3.86	−2.78	−2.63	−3.72	−2.36
	Std.	3.3×10^{-3}	4.03×10^{-6}	1.81×10^{-3}	3.89×10^{-1}	2.54	2.77	2.44
	Ranking	1	1	4	6	9	8	9
SCA	Average	4×10^{-1}	3	−3.85	−2.92	−2.9	−3.1	−4.03
	Std.	2.56×10^{-3}	9.16×10^{-5}	3×10^{-3}	2.72×10^{-1}	1.89	1.77	1.40
	*p*_value	2.9×10^{-1}	3.36×10^{-4}	8.21×10^{-3}	1.05×10^{-1}	6.44×10^{-1}	3.11×10^{-1}	1.89×10^{-3}
	Ranking	3	2	5	5	8	9	8
MRA	Average	4.26×10^{-1}	7.62	−3.7	−2.73	$-1.02 \times 10^{+1}$	$-1.04 \times 10^{+1}$	$-1.05 \times 10^{+1}$
	Std.	2.8×10^{-2}	4.46	7.3×10^{-2}	1.86×10^{-1}	2.85×10^{-3}	1.27×10^{-3}	4.9×10^{-3}
	*p*_value	2.77×10^{-6}	4.68×10^{-7}	3.53×10^{-17}	5.38×10^{-1}	3.08×10^{-23}	4.1×10^{-19}	7.97×10^{-26}
	Ranking	7	6	7	7	2	3	3
LCA	Average	7.7×10^{-1}	$2.35 \times 10^{+1}$	−3.21	−1.78	-5.02	−4.97	−4.45
	Std.	6.82×10^{-1}	$1.03 \times 10^{+1}$	4.2×10^{-1}	3.84×10^{-1}	9.63×10^{-1}	6.89×10^{-1}	1.59
	*p*_value	4.24×10^{-3}	1.23×10^{-15}	1.01×10^{-11}	3.73×10^{-14}	1.11×10^{-5}	1.95×10^{-2}	2.28×10^{-4}
	Ranking	8	8	8	8	7	7	7
CSA	Average	8.45×10^{-1}	$3.27 \times 10^{+1}$	−1.9	−1.17	$-1.02 \times 10^{+1}$	$-1.04 \times 10^{+1}$	$-1.05 \times 10^{+1}$
	Std.	8.82×10^{-16}	1.45×10^{-14}	6.78×10^{-16}	2.26×10^{-16}	3.61×10^{-15}	0	3.61×10^{-15}
	*p*_value	5.97×10^{-117}	0	2.12×10^{-169}	1.81×10^{-30}	3.05×10^{-23}	4.08×10^{-19}	7.88×10^{-26}
	Ranking	9	9	9	9	1	2	2
TSA	Average	3.99×10^{-1}	$1.23 \times 10^{+1}$	−3.86	−3.16	−7.15	−5.84	−4.72
	Std.	1.81×10^{-3}	$2.26 \times 10^{+1}$	3.25×10^{-3}	1.31×10^{-1}	1.27	2.17	2.77
	*p*_value	8.34×10^{-1}	2.86×10^{-2}	8.41×10^{-9}	3.15×10^{-6}	3.99×10^{-12}	1.69×10^{-3}	8.99×10^{-4}
	Ranking	2	7	3	4	5	6	6
DOA	Average	4×10^{-1}	3.9	−3.83	−3.25	−8.5	−7.97	−7.22
	Std.	1.27×10^{-2}	4.93	1.41×10^{-1}	$7.07 \times 10^{+2}$	2.63	2.78	3.58
	*p*_value	6.26×10^{-1}	3.21×10^{-1}	3.61×10^{-1}	1.31×10^{-8}	2.91×10^{-12}	1.76×10^{-7}	7.82×10^{-8}
	Ranking	6	4	6	3	4	5	5
EHO	Average	4×10^{-1}	3	−3.86	−3.27	−6.65	−8.15	−7.46
	Std.	2.65×10^{-4}	8.22×10^{-5}	2.71×10^{-15}	5.92×10^{-2}	3.45	3.29	3.86
	*p*_value	1.47×10^{-2}	4.11×10^{-4}	9.88×10^{-30}	3.43×10^{-09}	3.45×10^{-09}	5.17×10^{-07}	8.65×10^{-08}
	Ranking	3	2	1	2	6	4	4
WSO	Average	4×10^{-1}	3.9	−3.86	−3.3	−9.49	$-1.04 \times 10^{+1}$	$-1.05 \times 10^{+1}$
	Std.	2.71×10^{-4}	$3.12 \times 10^{+1}$	2.71×10^{-15}	4.84×10^{-02}	2.07	0.00	9.03×10^{-15}
	*p*_value	1.12×10^{-1}	3.25×10^{-1}	9.88×10^{-30}	8.54×10^{-10}	1.49×10^{-16}	4.08×10^{-19}	2.88×10^{-26}
	Ranking	3	4	1	1	3	1	1

Regarding statistical significance, the MRSO shows significant results in two out of the seven functions. For example, in F18, the MRSO has a *p*-value of 3.36×10^{-4}, indicating that its performance is statistically significant compared to most of the other algorithms, including the MRA and the WSO. In F19, the MRSO's *p*-value of 8.21×10^{-3} confirms that it performs significantly better than several algorithms, including the SCA and the DOA. However, in functions like F20 and F21, the MRSO does not achieve statistical significance, with *p*-values of 1.05×10^{-1} and 6.44×10^{-1}, respectively, suggesting that its performance is less conclusive in these cases. Overall, the MRSO demonstrates statistically significant improvements in two functions, while maintaining competitive performance across the rest. Table 5 highlights the MRSO's strengths in terms of average performance, stability, and statistical significance in the fixed-dimension multimodal functions. While it excels in several key areas, there are opportunities for further refinement in functions where its performance falls behind other metaheuristic algorithms.

To rank the algorithms, we used the Friedman mean ranking score [39], which measures how well each algorithm performed across all 23 classical benchmark functions. A lower score means better performance. As shown in Table 6, the MRSO received a ranking score of 10.2, placing it in the middle of the group. The MRA (7.5) and the CSA (8.1) performed better than the MRSO, taking the top spots. The TSA and the DOA followed closely with scores of 11.3 and 10.6, respectively. The SCA, the EHO, and the WSO ranked lower with scores of 12.8, 12.4, and 13.2, while the LCA had the lowest performance with a score of 15.1. These results show that while the MRSO did well in some benchmarks, especially in multimodal and fixed-dimension functions, the MRA and the CSA were stronger overall across all the tested functions.

Table 6. Scores comparison of MRSO, SCA, MRA, LCA, CSA, TSA, DOA, EHO, and WSO on 23 classical benchmark functions.

Algorithm	MRSO	SCA	MRA	LCA	CSA	TSA	DOA	EHO	WSO
Ranking	10.2	12.8	7.5	15.1	8.1	11.3	10.6	12.4	13.2

5.6. The MRSO Algorithm and Metaheuristic Approaches: Insights from the CEC-C06 2019 Benchmarking

As seen in Table 7, the MRSO outperformed eight metaheuristic algorithms in five out of the ten CEC-C06 2019 benchmark functions, particularly in functions F2, F3, F6, F7, and F10. For instance, the MRSO achieved an average of $1.83 \times 10^{+1}$ in F2, which was better than most other algorithms. In F6, the MRSO's average of $1.09 \times 10^{+1}$ also surpassed all other algorithms, demonstrating its ability to handle complex optimization problems effectively. However, the MRSO did not perform as well in F1 and F5, where its average was higher than some of the competing algorithms, indicating that there is room for improvement in certain benchmarks.

In terms of standard deviation, the MRSO demonstrated consistency in its performance across multiple functions. For example, in F2 and F3, the MRSO had very low standard deviations of 7.23×10^{-3} and 1.34×10^{-6}, respectively, indicating reliable and stable performance. The MRSO also performed consistently in F6 with a standard deviation of 1.01, which was lower than most other algorithms. However, in functions like F1, the MRSO had a higher standard deviation of $3.20 \times 10^{+5}$, showing greater variability in its results compared to other algorithms, which suggests that further refinement may be necessary to improve its stability across all benchmarks.

Table 7. Metrics-based comparison of MRSO and eight algorithms on CEC-C06 2019 benchmark functions.

Algorithm	Metric/Function	F1	F2	F3	F4	F5	F6	F7	F8	F9	F10
MRSO	Average	$1.59 \times 10^{+5}$	$1.83 \times 10^{+1}$	$1.37 \times 10^{+1}$	$9.20 \times 10^{+3}$	4.57	$1.09 \times 10^{+1}$	$6.11 \times 10^{+2}$	6.31	$4.97 \times 10^{+2}$	$2.13 \times 10^{+1}$
	Std.	$3.20 \times 10^{+5}$	7.23×10^{-3}	1.34×10^{-6}	$3.20 \times 10^{+3}$	4.13×10^{-1}	1.01	$2.29 \times 10^{+2}$	4.14×10^{-1}	$1.49 \times 10^{+2}$	1.49×10^{-1}
	Ranking	4	1	3	6	6	1	1	3	4	1
SCA	Average	$1.22 \times 10^{+10}$	$1.85 \times 10^{+1}$	$1.37 \times 10^{+1}$	$1.68 \times 10^{+3}$	3.22	$1.21 \times 10^{+1}$	$7.9 \times 10^{+2}$	6.06	$1.13 \times 10^{+2}$	$2.15 \times 10^{+1}$
	Std.	$1.79 \times 10^{+10}$	9.51×10^{-2}	1.05×10^{-4}	$7.12 \times 10^{+2}$	7.84×10^{-2}	6.76×10^{-1}	$1.47 \times 10^{+2}$	4.17×10^{-1}	$8.31 \times 10^{+1}$	7.81×10^{-2}
	p_value	4.31×10^{-4}	1.95×10^{-10}	2.74×10^{-9}	3.52×10^{-18}	5.34×10^{-25}	3.06×10^{-6}	6.55×10^{-4}	2.48×10^{-2}	8.56×10^{-18}	8.12×10^{-8}
	Ranking	9	3	4	2	2	4	2	2	3	4
MRA	Average	1.00	$1.58 \times 10^{+4}$	$1.37 \times 10^{+1}$	$2.53 \times 10^{+4}$	8.13	$1.35 \times 10^{+1}$	$2 \times 10^{+3}$	7.83	$4.39 \times 10^{+3}$	$2.17 \times 10^{+1}$
	Std.	0	$4.3 \times 10^{+3}$	9.52×10^{-4}	$8.93 \times 10^{+3}$	1.21	6.77×10^{-1}	$2.26 \times 10^{+2}$	3.83×10^{-1}	$7.03 \times 10^{+2}$	1.17×10^{-1}
	p_value	8.62×10^{-3}	8.77×10^{-28}	1.19×10^{-16}	4.84×10^{-13}	6.38×10^{-22}	8.34×10^{-17}	1.73×10^{-31}	2.29×10^{-21}	9.66×10^{-37}	1.05×10^{-16}
	Ranking	1	9	7	8	8	8	8	8	8	7
LCA	Average	$2.46 \times 10^{+8}$	$5.08 \times 10^{+1}$	$1.37 \times 10^{+1}$	$2.19 \times 10^{+4}$	6.74	$1.39 \times 10^{+1}$	$1.68 \times 10^{+3}$	7.25	$3.04 \times 10^{+3}$	$2.18 \times 10^{+1}$
	Std.	$3.16 \times 10^{+8}$	$3.26 \times 10^{+1}$	1.41×10^{-3}	7.83×10^{3}	1.04	7.35×10^{-1}	$2.94 \times 10^{+2}$	4.07×10^{-1}	$8.43 \times 10^{+2}$	1.37×10^{-1}
	p_value	7.56×10^{-5}	1.08×10^{-6}	2.75×10^{-21}	2.91×10^{-11}	3.61×10^{-15}	5.34×10^{-19}	1.48×10^{-22}	2.63×10^{-12}	3×10^{-23}	8.42×10^{-19}
	Ranking	7	8	8	7	7	9	7	7	7	8
CSA	Average	$6.48 \times 10^{+5}$	$1.95 \times 10^{+1}$	$1.37 \times 10^{+1}$	$4.41 \times 10^{+4}$	9.13	$1.25 \times 10^{+1}$	$2.1 \times 10^{+3}$	7.89	$4.71 \times 10^{+3}$	$2.19 \times 10^{+1}$
	Std.	5.44×10^{-10}	3.61×10^{-15}	9.03×10^{-15}	1.16×10^{-1}	3.85×10^{-7}	4.11×10^{-1}	5.88×10^{-2}	2.9×10^{-2}	4.15×10^{-5}	2.34×10^{-2}
	p_value	1.49×10^{-11}	4.3×10^{-120}	6×10^{-215}	8.9×10^{-54}	4.59×10^{-54}	2.94×10^{-10}	4.48×10^{-41}	1.18×10^{-28}	1.43×10^{-77}	4.5×10^{-30}
	Ranking	5	6	9	9	9	5	9	9	9	9
TSA	Average	$6.05 \times 10^{+4}$	$1.95 \times 10^{+1}$	$1.37 \times 10^{+1}$	$6.55 \times 10^{+3}$	4.31	$1.19 \times 10^{+1}$	$8.49 \times 10^{+2}$	6.55	$6.67 \times 10^{+2}$	$2.15 \times 10^{+1}$
	Std.	$1.66 \times 10^{+4}$	6.46×10^{-1}	1.5×10^{-3}	$4.21 \times 10^{+3}$	8.13×10^{-1}	8.1×10^{-1}	$2.43 \times 10^{+2}$	3.97×10^{-1}	$5.79 \times 10^{+2}$	1.01×10^{-1}
	p_value	9.8×10^{-2}	2.06×10^{-13}	2.04×10^{-3}	7.9×10^{-3}	1.15×10^{-1}	2.17×10^{-4}	2.55×10^{-4}	2.77×10^{-2}	1.24×10^{-1}	1.67×10^{-6}
	Ranking	2	7	6	5	5	2	4	4	6	3
DOA	Average	$1.16 \times 10^{+5}$	$1.84 \times 10^{+1}$	$1.37 \times 10^{+1}$	$4.37 \times 10^{+3}$	3.65	$1.32 \times 10^{+1}$	$1.02 \times 10^{+3}$	6.6	$5.62 \times 10^{+2}$	$2.16 \times 10^{+1}$
	Std.	$1.88 \times 10^{+5}$	1.89×10^{-1}	8.11×10^{-4}	$3.72 \times 10^{+3}$	6.37×10^{-1}	8.09×10^{-1}	$3.08 \times 10^{+2}$	3.66×10^{-1}	$3.17 \times 10^{+2}$	1.56×10^{-1}
	p_value	5.06×10^{-1}	7.13×10^{-2}	1.18×10^{-2}	1.33×10^{-6}	1.16×10^{-8}	1.22×10^{-13}	2.74×10^{-7}	5.48×10^{-3}	3.14×10^{-1}	7.1×10^{-12}
	Ranking	3	2	5	3	4	6	6	5	5	5
EHO	Average	$2.42 \times 10^{+9}$	$1.85 \times 10^{+1}$	$1.37 \times 10^{+1}$	$3.22 \times 10^{+1}$	3.22	$1.21 \times 10^{+1}$	$9.35 \times 10^{+2}$	6.74	3.62	$2.14 \times 10^{+1}$
	Std.	$4.12 \times 10^{+9}$	9.67×10^{-2}	9.03×10^{-15}	$1.66 \times 10^{+1}$	1.52×10^{-2}	2.71	$4.71 \times 10^{+2}$	6.06×10^{-1}	3.52×10^{-1}	2.08×10^{-1}
	p_value	2.13×10^{-3}	1.95×10^{-11}	1.33×10^{-3}	1.57×10^{-22}	3.11×10^{-18}	3.98×10^{-2}	1.25×10^{-3}	2.28×10^{-3}	1.62×10^{-25}	4.78×10^{-2}
	Ranking	8	3	1	1	2	3	5	6	1	2
WSO	Average	$2.92 \times 10^{+7}$	$1.85 \times 10^{+1}$	$1.37 \times 10^{+1}$	$4.37 \times 10^{+3}$	2.27	$1.32 \times 10^{+1}$	$8.16 \times 10^{+2}$	5.5	$3.62 \times 10^{+1}$	$2.16 \times 10^{+1}$
	Std.	$6.88 \times 10^{+7}$	7.66×10^{-2}	8.23×10^{-11}	$4.11 \times 10^{+3}$	2.29×10^{-1}	7.41×10^{-1}	$1.58 \times 10^{+2}$	6.61×10^{-1}	$1.04 \times 10^{+2}$	1.56×10^{-1}
	p_value	2.43×10^{-2}	2.71×10^{-13}	5.9×10^{-4}	2.57×10^{-8}	2.92×10^{-34}	2.11×10^{-15}	1.69×10^{-4}	3.93×10^{-7}	4.85×10^{-20}	7.1×10^{-12}
	Ranking	6	3	1	3	1	6	3	1	2	5

The p-values, as shown in Table 7, further highlight the MRSO's statistical significance in several functions. In F2, the MRSO's p-value of 1.95×10^{-10} indicates a statistically significant improvement over competing algorithms, confirming the robustness of its performance. Similarly, in F6, the p-value of 3.06×10^{-6} demonstrates that the MRSO's results are not due to random chance and are significantly better than other algorithms. On the other hand, in functions like F5, the p-value of 5.34×10^{-25} shows that other algorithms, like the WSO, performed better than the MRSO. Despite this, the MRSO consistently showed significant results in key functions, reinforcing its competitiveness. Table 7 highlights the MRSO's strengths in handling diverse optimization problems, showing strong performance in several key benchmarks against eight other metaheuristic algorithms, with statistically significant improvements in many areas.

To systematically rank the algorithms, we again used the Friedman mean ranking score to evaluate the overall performance of each algorithm across all ten CEC-C06 2019 benchmark functions. As shown in Table 8, the MRSO achieved a ranking score of 3, placing it among the top performers. The SCA followed closely with a score of 3.5, and the TSA and the DOA both ranked similarly with a score of 4.4. The WSO had a slightly better score of 3.1, indicating strong performance. On the other hand, the MRA, the LCA, and the CSA ranked lower, with scores of 7.2, 7.5, and 7.9, respectively, showing that these algorithms were less effective overall compared to the MRSO and the other top-ranked algorithms. These rankings highlight the MRSO's strong capability in handling the CEC-C06 2019 benchmark functions, outperforming most of the competing algorithms.

Table 8. Scores comparison of MRSO, SCA, MRA, LCA, CSA, TSA, DOA, EHO, and WSO on CEC-C06 2019 benchmark functions.

Algorithm	MRSO	SCA	MRA	LCA	CSA	TSA	DOA	EHO	WSO
Ranking	3	3.5	7.2	7.5	7.9	4.4	4.4	3.2	3.1

At the conclusion of Sections 5.5 and 5.6, our proposed MRSO algorithm performs better on the 10 CEC-C06 2019 benchmark functions than on the classical benchmark functions. This suggests that the MRSO is particularly effective at handling complex, multidimensional optimization problems, demonstrating its strength in addressing more challenging tasks compared to other metaheuristic algorithms. These results confirm the algorithm's potential for broader applications in solving difficult optimization challenges.

6. Utilizing MRSO for Solving Engineering Design Issues

This section explores the application of the Modified Rat Swarm Optimizer (MRSO) for tackling various engineering design problems. It provides a theoretical overview of seven real-world constrained engineering design challenges and compares the performance of the MRSO with the original Rat Swarm Optimizer (RSO). Through this comparison, the effectiveness and improvements offered by the MRSO in solving these complex design issues are highlighted.

6.1. Overview of Constrained Engineering Design Problems

In this subsection, we present a theoretical description of seven significant engineering design problems that are commonly encountered in practice. These problems include Pressure Vessel Design, Tension/Compression Spring Design, Three Bar Truss, Gear Train Design, Cantilever Beam, and Welded Beam. Each problem is characterized by its unique set of constraints and objectives, demonstrating the complexity and diversity of engineering optimization tasks.

6.1.1. Pressure Vessel Design

Designing a pressure vessel, as proposed by Kannan and Kramer (1994) [40], aims to minimize the overall cost, which includes material, shaping, and welding. The vessel, shown in Figure 3, consists of a cylindrical body with hemispherical heads at both ends. The problem involves four design variables:

S_t = shell thickness
H_t = head thickness
I_r = inner radius
L_c = length of the cylindrical section

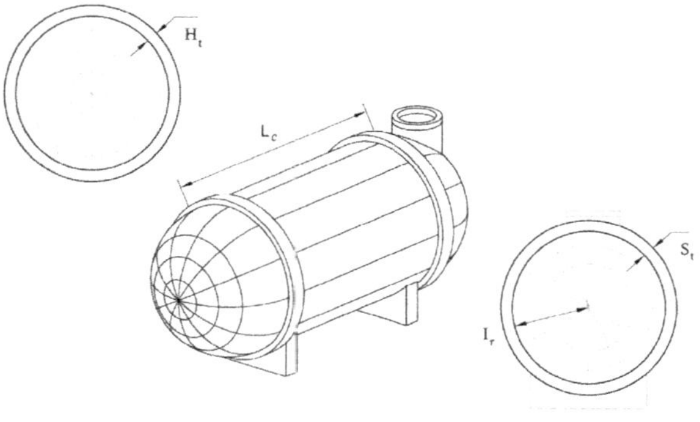

Figure 3. Pressure Vessel Design diagram.

Here, I_r and L_c are continuous variables, while S_t *and* H_t are discrete values that are multiples of 0.0625 inches.

Consider the following:

$$\vec{P} = [P_1, P_2, P_3, P_4] = [S_t, H_t, I_r, L_c]$$

The objective is to minimize the cost function:

$$f\left(\vec{P}\right) = 0.6224P_1P_3P_4 + 1.7781P_2P_3^2 + 3.1661P_1^2P_4 + 19.84P_1^2P_3$$

subject to the following constraints:

$$S_1\left(\vec{P}\right) = -P_1 + 0.0193P_3 \leq 0$$
$$S_2\left(\vec{P}\right) = -P_3 + 0.00954P_3 \leq 0$$
$$S_3\left(\vec{P}\right) = -\pi P_3^2P_4 - \tfrac{4}{3}\pi P_3^3 + 1{,}296{,}000 \leq 0$$
$$S_4\left(\vec{P}\right) = -P_4 - 240 \leq 0$$

with variable ranges:

$$0 \leq P_1 \leq 99,\ 0 \leq P_2 \leq 99,\ 10 \leq P_3 \leq 200,\ 10 \leq P_4 \leq 200$$

6.1.2. Tension/Compression Spring Design (TCSD)

The TCSD problem, shown in Figure 4, seeks to minimize the volume of a coil spring under constant load. It involves three design variables [41,42]:

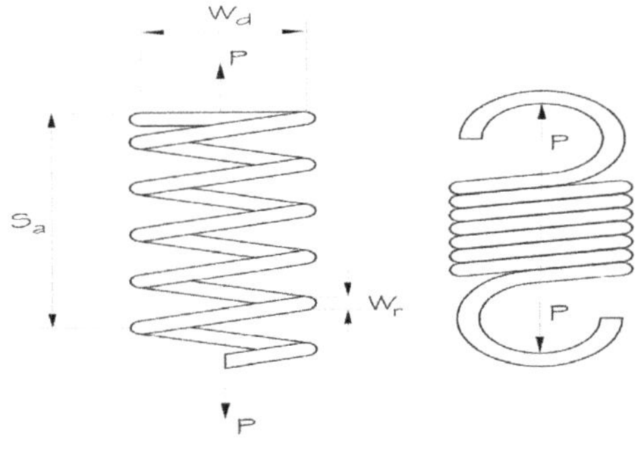

Figure 4. Diagram of the Tension/Compression Spring Design.

Consider the following:

$$\vec{P} = [P_1,\ P_2,\ P_3] = \left[W_r\ (\text{wire diameter}),\ W_d\ (\text{winding diameter}),\ S_a(\text{number of spring's activecoils})\right]$$

The objective is to minimize the cost function:

$$f\left(\vec{P}\right) = (P_3 + 2)P_2P_1^2$$

Subject to the following:

$$S_1\left(\vec{P}\right) = 1 - \frac{P_2^3}{71785P_1^4} \leq 0$$
$$S_2\left(\vec{P}\right) = \frac{4P_2^2 - P_1P_2}{12566\left(P_2P_1^3 - P_1^4\right)} + \frac{1}{5108P_1^2} - 1 \leq 0$$
$$S_3\left(\vec{P}\right) = 1 - \frac{140.45P_1}{P_2^2P_3} \leq 0$$
$$S_4\left(\vec{P}\right) = -\frac{P_1 + P_2}{1.5} - 1 \leq 0$$

with variable bounds:

$$0.05 \leq P_1 \leq 2.00$$
$$0.25 \leq P_2 \leq 1.30$$
$$2.00 \leq P_3 \leq 15.00$$

6.1.3. Three Bar Truss Design

Introduced by Ray and Saini [43,44], this problem aims to minimize the weight of a Three Bar Truss while considering stress, deflection, and buckling constraints. The design variables are the cross-sectional areas of the bars. The desired placement of the bars is shown in Figure 5. The objective is to minimize the cost function [44]:

$$f(B_1,\ B_2) = \left(2\sqrt{2B_1} + B_2\right) \times l$$

subject to the following:

$$S_1 = \frac{\sqrt{2}B_1 + B_2}{\sqrt{2B_1^2 + 2B_1B_2}}P - P \leq 0$$
$$S_2 = \frac{B_2}{\sqrt{2B_1^2 + 2B_1B_2}}P - P \leq 0$$
$$S_3 = \frac{1}{\sqrt{2B_1^2 + 2B_1B_2}}P - P \leq 0$$

where

$$0 \le B_1,\ B_2 \le 1,\ l = 100\ \text{cm}\ and\ P = 2\frac{\text{KN}}{\text{cm}^2}\ (kilonewton/square\ centimetre)$$

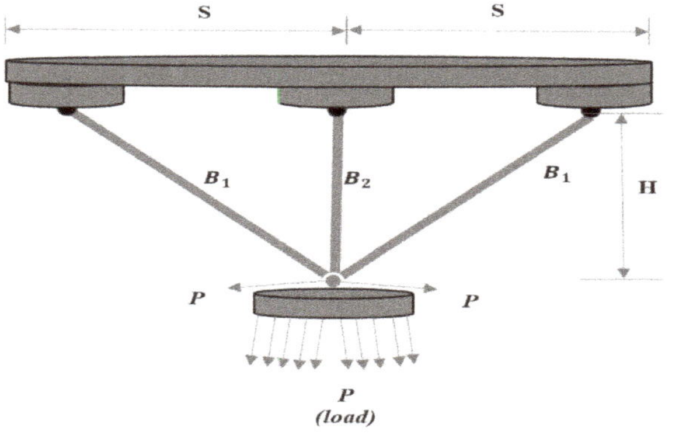

Figure 5. Diagram of the Three Bar Truss problem.

6.1.4. Gear Train Design

The Gear Train Design problem is a type of unconstrained optimization involving four integer variables. Initially introduced by Sandgran [45], this problem focuses on minimizing the cost associated with the gear ratio of a specific gear train, illustrated in Figure 6. The gear ratio is calculated using the following formula [46]:

$$Gear\ ratio = \frac{Gt_a \times Gt_b}{Gt_c \times Gt_d}$$

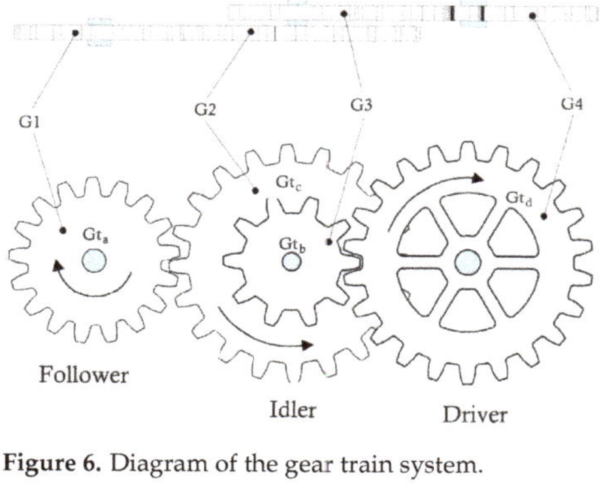

Figure 6. Diagram of the gear train system.

Here, Gt_a, Gt_b, Gt_c, and Gt_d represent the number of teeth on gears G_1, G_2, G_3, and G_4, respectively. Each Gt_i indicates the number of teeth on the gearwheel G_i. The mathematical expression for this optimization problem is as follows:

$$f(Gt_a, Gt_b, Gt_c, Gt_d) = \left(\frac{1}{6.931} - \frac{Gt_a \times Gt_b}{Gt_c \times Gt_d}\right)$$

This formula is aimed at optimizing the gear ratio to achieve cost efficiency in the design [46].

6.1.5. Cantilever Beam Design

This problem involves optimizing the weight of a Cantilever Beam with a square cross-section, as shown in Figure 7. The beam is fixed at one end and subjected to a vertical force at the other. The design variables are the heights (or widths) of the beam elements, while the thickness is fixed (t = 2/3). These constraints ensure that the beam can carry the applied load without failing. The objective is to minimize the following [46]:

$$f(x) = 0.0624(x_1 + x_2 + x_3 + x_4 + x_5)$$

subject to the following:

$$S(x) = \frac{61}{x_1^3} + \frac{37}{x_2^3} + \frac{19}{x_3^3} + \frac{7}{x_4^3} + \frac{1}{x_5^3} - 1 \leq 0$$

with constraints on the following variable bounds: $0.01 \leq X_J \leq 100$.

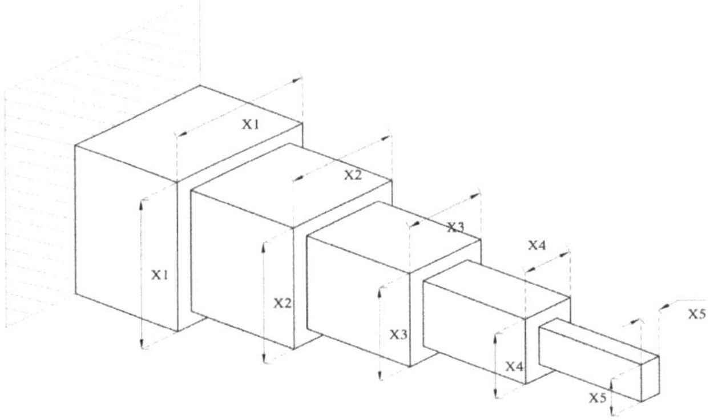

Figure 7. Diagram of the Cantilever Beam issue.

6.1.6. Welded Beam Design

The goal here is to minimize the cost of producing a welded beam, as illustrated in Figure 8. Subject to several constraints ensuring structural integrity and performance, these constraints include limits on stress, deflection, and stability of the welded beam under the applied loads.

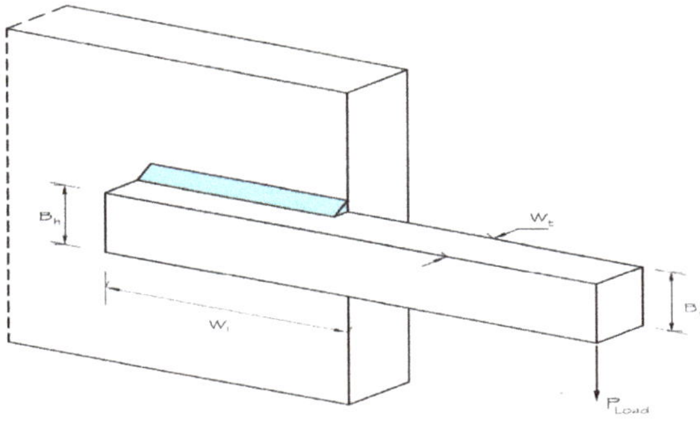

Figure 8. Welded Beam Design schematic diagram.

The design variables are as follows [8,42]:

Consider the following:

$$\vec{P} = [W_t,\ W_l,\ B_h, B_t\ ] = [weld\ thickness,\ length\ of\ the\ welded\ section,\ bar\ height, bar\ thickness\ ]$$

The objective is to minimize the following:

$$f\left(\vec{P}\right) = 1.10471W_t^2W_l + 0.04811B_hB_t(14.0 + W_l)$$

subject to the following:

$$S_1\left(\vec{P}\right) = \tau\left(\vec{P}\right) - \tau_{max} \leq 0$$
$$S_2\left(\vec{P}\right) = \sigma\left(\vec{P}\right) - \sigma_{max} \leq 0$$
$$S_3\left(\vec{P}\right) = \delta\left(\vec{P}\right) - \delta_{max} \leq 0$$
$$S_4\left(\vec{P}\right) = W_t - B_t \leq 0$$
$$S_5\left(\vec{P}\right) = P - P_c\left(\vec{P}\right) \leq 0$$
$$S_6\left(\vec{P}\right) = 0.125 - W_t \leq 0$$
$$S_7\left(\vec{P}\right) = 1.10471PW_t^2 + 0.04811B_hB_t(14.0 + W_l) - 5.0 \leq 0$$

with constraints on the following variable bounds: $0.05 \leq W_t \leq 2.00,\ 0.25 \leq W_l \leq 1.30$ *and* $2.00 \leq B_h,\ B_t \leq 15.00$.

6.1.7. Tire Design Problem

The Tire Design problem focuses on minimizing tire deflection, which is critical to improving vehicle performance, safety, and longevity. Tire deflection occurs when a tire deforms under load, impacting factors such as handling, fuel efficiency, and wear. The optimization involves key parameters, including the aspect ratio, section width, load capacity, pressure, and rim diameter, which influence the tire's stiffness and ability to maintain its shape under varying loads and road conditions. By adjusting these parameters, engineers aim to strike a balance between minimizing deflection and maintaining comfort, traction, and durability [47,48]. This problem, as illustrated in Figure 9, is common in vehicle dynamics optimization and is often addressed using techniques such as constrained nonlinear optimization.

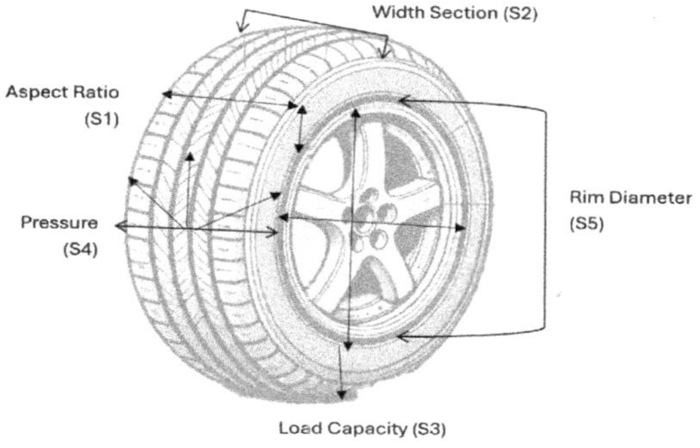

Figure 9. Tire Design problem schematic diagram.

The design variables are as follows:
Consider the following:

$$\vec{P} = [S_1, S_2, S_3, S_4, S_5] = [Aspect\ Ratio,\ Width\ Section,\ load\ Capacity,\ Pressure, Rim\ Diameter\ ]$$

The objective is to minimize the following:

$$f\left(\vec{P}\right) = F_z \times \frac{W_1 + W_2 + W_3}{K_1}$$

subject to the following:

$$R = \frac{S_5 \times 25.4}{2}$$

$$H = \frac{S_2 \times S_1}{100}$$

$$T = \frac{\left(\frac{H}{2}\right) \times \left(1 - \left(\frac{S_1}{100}\right)\right)}{1000}$$

$$I = \left(\frac{H^3}{12}\right) \times T \times 1{,}000{,}000$$

$$F_z = \frac{S_3}{4}$$

$$P = S_4 \times 1000$$

$$A = \sqrt{(R^2 - (\frac{H}{2})^2)}$$

$$B = \sqrt{(R^2 - (T + \frac{H}{2})^2)}$$

$$W_1 = K_1 \times \left(\frac{F_z}{P}\right)^{0.9}$$

$$W_1 = K_2 \times \left(\frac{F_z}{P}\right)^{0.8} \times \frac{A}{B}^{1.1}$$

$$W_1 = K_3 \times \left(\frac{F_z}{P}\right)^{0.7} \times \frac{A}{B}^{1.3} \times \frac{I}{H^3}^{0.7}$$

where $K_1 = 5000$, $K_2 = 10{,}000$, and $K_3 = 20{,}000$.

6.2. Comparative Analysis of MRSO and RSO in Engineering Applications

This subsection presents a comparative analysis of the MRSO versus the original RSO across seven engineering design problems. The objective is to emphasize the enhancements and benefits of the MRSO in solving constrained engineering design problems more efficiently. The comparison highlights the practical improvements brought by the MRSO in terms of performance and reliability.

The results from Table 9 show that the MRSO consistently outperforms the original RSO in terms of average performance across all seven engineering design problems [49]. The average performance metrics in Table 9 clearly demonstrate the MRSO's superiority over the RSO. In the Pressure Vessel Design problem, the MRSO achieves an average cost of $6.956 \times 10^{+3}$, which is significantly better than the RSO's $1.720 \times 10^{+4}$. Similarly, in the String Design problem, the MRSO vastly outperforms the RSO, achieving an average cost of 1.417×10^{-2} compared to the RSO's $3.954 \times 10^{+8}$. These dramatic improvements indicate the MRSO's ability to find more optimal solutions across various complex design challenges. The Gear Train Design and Cantilever Beam problems also reflect this trend, with the MRSO obtaining much lower average values (1.106×10^{-12} and 1.343, respectively) compared to the RSO's higher costs (4.792×10^{-3} and 2.382). Across all seven problems, the MRSO provides more efficient solutions, outclassing the RSO in terms of lower average values in every case.

Table 9. The MRSO vs. standard RSO performance when applied to the seven engineering design problems: average, standard deviation, and statistical significance.

Engineering Application	MRSO		RSO	
	Avg.	Std.	Avg.	Std.
Pressure Vessel Design	$6.956 \times 10^{+3}$	$4.479 \times 10^{+2}$	$1.72 \times 10^{+4}$	$1.031 \times 10^{+4}$
String Design	1.417×10^{-2}	9.721×10^{-4}	$3.954 \times 10^{+8}$	$4.217 \times 10^{+8}$
Three Bar Truss	$2.665 \times 10^{+2}$	6.522	$2.705 \times 10^{+2}$	4.829
Gear Train Design	1.106×10^{-12}	2.739×10^{-12}	4.792×10^{-3}	6.392×10^{-3}
Cantilever Beam	1.343	2.932×10^{-3}	2.382	8.673×10^{-1}
Welded Beam	1.552	2.052×10^{-1}	$7.253 \times 10^{+7}$	$1.629 \times 10^{+8}$
Tire Design	4.683×10^{-3}	1.764×10^{-18}	1.74	1.162

In terms of consistency, reflected by standard deviation, the MRSO also demonstrates superior performance in most cases. For instance, in the Pressure Vessel Design, the MRSO shows a significantly lower standard deviation ($4.479 \times 10^{+2}$) compared to the RSO ($1.031 \times 10^{+4}$), indicating that the MRSO is not only more efficient but also more stable across iterations. In the Gear Train Design, the MRSO exhibits much smaller variability with a standard deviation of 2.739×10^{-12}, compared to the RSO's higher variation of 6.392×10^{-3}. However, there are instances where the RSO shows slightly better consistency, such as in the Three Bar Truss problem, where the RSO achieves a lower standard deviation (4.829) compared to the MRSO (6.522). This suggests that while the MRSO offers substantial improvements in most cases, there are still some applications where further refinement could help improve its consistency.

While the MRSO demonstrates significant improvements over the RSO, there are still cases where both algorithms exhibit weaknesses, particularly in handling complex constraints or highly nonlinear functions. For example, in the Three Bar Truss problem, despite the MRSO having a slightly better average value, both algorithms show relatively close performance with only marginal differences. This suggests that both the MRSO and the RSO may face challenges in solving certain structural design problems with high dimensionality or multiple constraints. Additionally, in the Welded Beam problem, while the MRSO outperforms the RSO, the higher standard deviation for both algorithms indicate instability, suggesting that further refinement may be necessary to improve convergence and reliability in these types of problems. These failure points highlight the need for the continued optimization of the MRSO's parameters and potential hybridization with other algorithms to ensure robust performance across all types of engineering design applications.

6.3. Performance Comparison of MRSO with SCA, LCA, TSA, and DOA on Seven Engineering Design Problems

In this section, we compare the performance of the MRSO with four other metaheuristic algorithms—the SCA, the LCA, the TSA, and the DOA—across seven engineering design problems. These include the Pressure Vessel Design, String Design, Three Bar Truss, Gear Train Design, Cantilever Beam, Welded Beam, and Tire Design. Table 10 provides a detailed breakdown of the average, standard deviation, and ranking for each algorithm on these problems. Our evaluation was conducted by running each algorithm 30 times to ensure reliable and accurate results.

Table 10. Performance comparison of MRSO, SCA, LCA, TSA, and DOA on engineering design problems.

Engineering Problems	Metrics	Algorithms				
		MRSO	SCA	LCA	TSA	DOA
Pressure Vessel Design	Avg.	$6.83 \times 10^{+3}$	$7.28 \times 10^{+3}$	$1.64 \times 10^{+4}$	$7.24 \times 10^{+3}$	$5.98 \times 10^{+3}$
	Std.	$5.21 \times 10^{+2}$	$7.80 \times 10^{+2}$	$2.02 \times 10^{+3}$	$7.18 \times 10^{+2}$	$3.32 \times 10^{+3}$
	Rank	2	4	5	3	1
String Design	Avg.	1.38×10^{-2}	1.31×10^{-2}	$2.86 \times 10^{+8}$	1.45×10^{-2}	$6.37 \times 10^{+7}$
	Std.	8.98×10^{-4}	2.42×10^{-4}	$3.19 \times 10^{+8}$	1.72×10^{-3}	$1.65 \times 10^{+8}$
	Rank	2	1	5	3	4
Three Bar Truss	Avg.	$2.51 \times 10^{+2}$	$2.67 \times 10^{+2}$	$2.79 \times 10^{+2}$	$2.64 \times 10^{+2}$	$2.64 \times 10^{+2}$
	Std.	$6.12 \times 10^{+1}$	6.44	$1.13 \times 10^{+1}$	3.53×10^{-1}	3.59×10^{-1}
	Rank	1	4	5	3	2
Gear Train Design	Avg.	1.75×10^{-13}	2.01×10^{-9}	5.28×10^{-3}	5.44×10^{-10}	1.27×10^{-13}
	Std.	2.08×10^{-13}	4.63×10^{-9}	7.29×10^{-3}	9.02×10^{-10}	5.87×10^{-13}
	Rank	2	4	5	3	1
Cantilever Beam	Avg.	1.34	1.41	1.57	1.36	1.4
	Std.	2.93×10^{-3}	2.75×10^{-2}	4.63×10^{-2}	1.13×10^{-2}	9.62×10^{-2}
	Rank	1	4	5	2	3
Welded Beam	Avg.	1.55	1.59	$4.94 \times 10^{+7}$	1.56	1.95
	Std.	2.05×10^{-1}	4×10^{-2}	$1.36 \times 10^{+8}$	3.64×10^{-2}	6.13×10^{-1}
	Rank	1	3	5	2	4
Tire Design	Avg.	4.68×10^{-3}	4.68×10^{-3}	2.15	4.68×10^{-3}	8.24×10^{-3}
	Std.	1.76×10^{-18}	1.11×10^{-17}	1.31	8.82×10^{-19}	1.36×10^{-2}
	Rank	1	2	5	3	4

In terms of average values, the MRSO showed superior performance in four out of the seven engineering design problems. In the Pressure Vessel Design, the MRSO secured second place with an average of $6.83 \times 10^{+3}$, coming close to the DOA, which achieved the best performance with $5.98 \times 10^{+3}$. However, in the Three Bar Truss, the MRSO outperformed all other algorithms, achieving the lowest average value of $2.51 \times 10^{+2}$. Similarly, in the Cantilever Beam, the MRSO achieved the best average value of 1.34, proving its robustness in solving this particular problem. On the other hand, the MRSO ranked second in String Design, where the SCA led with an average of 1.31×10^{-2}, compared to the MRSO's 1.38×10^{-2}. In Gear Train Design, the MRSO was narrowly beaten by the DOA, placing second with an average of 1.75×10^{-13}. In Tire Design, the MRSO also took first place, highlighting its strength across multiple problems.

When considering standard deviation, the MRSO displayed strong consistency across most problems. In Pressure Vessel Design, the MRSO's standard deviation of $5.21 \times 10^{+2}$ was lower than the SCA and the LCA, indicating more reliable performance. For the Three Bar Truss, the MRSO's standard deviation of $6.12 \times 10^{+1}$ was considerably higher than the DOA and the TSA, suggesting some variability in its results, despite having the best average value. The MRSO also demonstrated stable results in the Cantilever Beam, with a lower standard deviation of 2.93×10^{-3}, compared to the TSA's 1.13×10^{-2} and the LCA's 4.63×10^{-2}. This highlights the reliability of the MRSO's performance in these cases.

Regarding consistency, as represented by standard deviation, the MRSO exhibited reliable performance across most of the engineering problems. For instance, in Pressure Vessel Design, the MRSO had a significantly lower standard deviation ($5.21 \times 10^{+2}$) than the SCA and the LCA, indicating more stable results across iterations. Similarly, the MRSO demonstrated strong consistency in the Cantilever Beam, where its standard deviation was 2.93×10^{-3}, considerably lower than the TSA (1.13×10^{-2}) and the LCA (4.63×10^{-2}). In Gear Train Design, the MRSO maintained a lower standard deviation (2.08×10^{-13})

compared to the SCA and the LCA, showcasing its reliability in delivering consistent results. However, in the Three Bar Truss, the MRSO exhibited a higher standard deviation of $6.12 \times 10^{+1}$, indicating some variability in its performance despite having the best average value.

When analyzing the ranking based on performance across the seven engineering problems, the MRSO secured top positions in four out of seven problems. In the Three Bar Truss and the Cantilever Beam, the MRSO ranked first, outperforming all other algorithms in these tasks. In the Welded Beam and Tire Design, the MRSO also achieved the best ranking, further highlighting its robustness. However, in Pressure Vessel Design and Gear Train Design, the MRSO ranked second, being narrowly outperformed by the DOA in both cases. In String Design, the MRSO came in second again, slightly behind the SCA. Overall, the MRSO demonstrated strong competitive performance, securing first or second place in six of the seven problems, as shown in Table 10.

The convergence rates, shown in Figure 10, indicate that the MRSO converges more quickly than the other algorithms in most problems, particularly in Pressure Vessel Design, Three Bar Truss, and Cantilever Beam. This rapid convergence showcases the MRSO's efficiency in finding optimal solutions early in the iteration process. However, the graph also reveals that for certain problems, such as the String Design and Gear Train Design, the MRSO struggles to outperform other algorithms like the DOA and the SCA, where the DOA's performance fluctuates but often outpaces the MRSO in terms of final optimal values. These findings suggest that while the MRSO is highly effective in most cases, there are specific areas, such as high-dimensional or more complex designs, where further tuning may be required to enhance its optimization capability.

Figure 10. Convergence rate comparison of MRSO and metaheuristic algorithms across engineering design problems.

A detailed analysis of the failures reveals that the MRSO and some of the competing algorithm's struggle with high-dimensional problems like String Design and Gear Train Design. In the case of String Design, both the MRSO and the DOA produce suboptimal solutions compared to the SCA, showing that the MRSO's exploration capabilities might need improvement for problems with numerous variables. Similarly, in Gear Train Design, while the MRSO performs better than most algorithms, it fails to achieve the best performance, falling behind the DOA. These failures provide valuable insights into the MRSO's limitations, particularly in handling more complex optimization landscapes. These problems could be fixed by making the algorithm better and mixing it with other metaheuristic methods. This would make it better at high-dimensional or non-convex optimization tasks.

7. Conclusions, Challenges, and Future Research

This section draws conclusions from the study, highlights the challenges encountered, and suggests areas for future research efforts.

7.1. Conclusions

In this research, we introduced the Modified Rat Swarm Optimizer (MRSO) as an improvement over the original Rat Swarm Optimizer (RSO), specifically designed to address its limitations in handling convergence and exploration. The MRSO incorporates key modifications aimed at achieving a better balance between exploration and exploitation, which are critical for solving complex optimization problems effectively.

We thoroughly tested the MRSO against eight recent and commonly used metaheuristic algorithms, including the SCA, the MRA, the LCA, the CSA, the TSA, the DOA, the EHO, and the WSO, on both classical benchmark functions and CEC 2019 benchmark functions. The results consistently demonstrated the MRSO's superiority in avoiding local optima and yielding better results than the standard RSO across a wide range of functions. The MRSO's competitive performance, particularly in handling multimodal and high-dimensional problems, highlights its robustness and effectiveness as an optimization tool.

Additionally, we applied the MRSO to seven real-world engineering design problems—Pressure Vessel Design, String Design, Three Bar Truss, Gear Train Design, Cantilever Beam, Welded Beam, and Tire Design. For this, we compared the MRSO's performance with four of the algorithms (the SCA, the LCA, the TSA, and the DOA). The results indicated that the MRSO not only outperformed the RSO but also consistently outperformed the other four metaheuristic algorithms in most cases. These findings emphasize the MRSO's reliability and efficiency in solving real-world engineering applications, where striking the right balance between exploration and exploitation is crucial to finding optimal solutions.

7.2. Limitations

Despite these promising results, there are notable limitations that must be addressed. First, our evaluation was confined to a specific set of benchmark functions and engineering problems. Although these problems are widely accepted in the optimization field, they may not represent the full spectrum of real-world challenges. Therefore, the results may not fully generalize to all problem types or industries, particularly for those that deviate from the tested scenarios.

Second, the parameter settings employed in our tests were fixed after initial tuning, which worked well under the tested conditions. However, it is conceivable that further fine-tuning or employing adaptive parameter control mechanisms could yield even better results. This underscores the need for a more detailed investigation into dynamic parameter settings that could adapt to the unique demands of different optimization problems.

Additionally, the scope of this study did not include testing the MRSO on large-scale, highly complex engineering problems, which may limit its immediate application to such contexts. Addressing large-scale optimization scenarios will be crucial for future research to determine how scalable the MRSO is when dealing with computationally intensive problems. Lastly, the comparative algorithms used in our evaluation, while competitive, were not exhaustive. A more comprehensive comparison involving newer or more established algorithms could provide deeper insights into the MRSO's performance advantages and its areas of improvement.

7.3. Future Work

In the future, research should focus on extending the MRSO's application to a wider range of optimization challenges, particularly those involving large-scale, real-world engineering problems. Doing so would help confirm how adaptable and scalable the MRSO is when faced with more complex, computationally demanding tasks, providing deeper insights into how well it can generalize beyond the current tests.

Another exciting area for future work is developing adaptive mechanisms that allow the MRSO to adjust its parameters automatically, depending on the specific problem it is tackling. This would not only make the MRSO more flexible but would also reduce the need for manual tuning, making it an even more powerful tool across a broader variety of optimization tasks.

Additionally, incorporating surrogate models or other techniques to lower computational costs could make the MRSO more efficient in solving high-dimensional problems. Since real-world scenarios often require a balance between computational resources and solution quality, integrating methods that improve the MRSO's efficiency could make it more practical for large-scale industry applications. Lastly, future studies should include a wider variety of metaheuristic algorithms and configurations, which will provide a clearer understanding of the MRSO's strengths and highlight areas where it could be further improved.

Author Contributions: Conceptualization, A.A.A.; methodology, A.A.A. and H.S.A.; software, A.A.A., H.S.A., S.I.S., I.A.M., and T.A.R.; validation, A.A.A. and T.A.R.; formal analysis, S.I.S. and I.A.M.; investigation, A.A A.; resources, A.A.A., H.S.A., S.I.S., I.A.M., and T.A.R.; data curation, A.A.A.; writing—original draft preparation, A.A.A.; writing—review and editing, H.S.A., S.I.S., I.A.M., and T.A.R.; visualization, A.A.A.; supervision, A.A.A.; project administration, A.A.A.; funding acquisition, H.S.A., S.I.S., and I.A.M. All authors have read and agreed to the published version of the manuscript.

Funding: This research received no external funding.

Data Availability Statement: The MATLAB code for the Modified Rat Swarm Optimization (MRSO) algorithm used in this study is available at (https://www.mathworks.com/matlabcentral/fileexchange/172930-modified-rat-swarm-optimization-mrso-algorithm).

Conflicts of Interest: The authors declare no conflicts of interest.

Appendix A

Tables A1–A3 present the mathematical formulations of the 23 standard benchmark functions used in this study. These tables provide a detailed overview of the equations, problem dimensions, upper and lower boundary ranges, and the minimum goal values for each function. This comprehensive information serves as the foundation for evaluating the performance of the algorithms tested [24,35].

Table A1. Unimodal test functions [24].

Function	Dimension	Range	f_{min}
$F_1(x) = \sum_{i=1}^{n} X_i^2$	10	[−100, 100]	0
$F_2(x) = \sum_{i=1}^{n} \lvert X_i \rvert + \prod_{i=1}^{n} \lvert X_i \rvert$	10	[−10, 10]	0

Table A1. *Cont.*

Function	Dimension	Range	f_{min}
$F_3(x) = \sum_{i=1}^{n}\left(\sum_{j=1}^{i} X_j\right)^2$	10	[−30, 30]	0
$F_4(x) = max_i\{\lvert X_i\rvert, 1 \le i \le n\}$	10	[−100, 100]	0
$F_5(x) = \sum_{i=1}^{n-1}\left[100\left(x_{i+1} - x_1^2\right)^2 + (x_i - 1)^2\right]$	10	[−30, 30]	0
$F_6(x) = \sum_{i=1}^{n}([x_i + 0.5])^2$	10	[−100, 100]	0
$F_7(x) = \sum_{i=1}^{n} ix_i^4 + random[0, 1]$	10	[−1.28, 1.28]	0

Table A2. Multimodal test functions [35].

Function	Dimension	Range	f_{min}
$F_8(x) = \sum_{i=1}^{n} -X_i \sin\left(\sqrt{\lvert x_i\rvert}\right)$	10	[−500, 500]	$-418.9829 \times n$ When n equals to dimensions
$F_9(x) = \sum_{i=1}^{n}\left[x_i^2 - 10\cos(2\pi x_i) + 10\right]$	10	[−10, 10]	0
$F_{10}(x) = -20exp\left(\sqrt[-0.2]{\sum_{i=1}^{n} x_i^2}\right) - exp\left(\frac{1}{n}\sum_{i=1}^{n}\cos(2\pi x_i)\right) + 20 + e$	10	[−32, 32]	0
$F_{11}(x) = \frac{1}{4000}\sum_{i=1}^{n} x_i^2 - \prod_{i=1}^{n}\cos\left(\frac{x_i}{\sqrt{i}}\right) + 1$	10	[−600, 600]	0
$F_{12}(x) = \frac{\pi}{n}\left\{10\sin(\pi y_1) + \sum_{i=1}^{n-1}(y_i - 1)^2[1 + 10\sin(\pi y_{i+1})^2] + (y_n - 1)^2\right\} + \sum_{i=1}^{n}\mu(x_i, 10, 100, 4),\ y_i = 1 + \frac{x+1}{4},\ \mu(x_i, a, k, m) = \begin{cases} k(x_i - a)^m\ x_i > a \\ 0 - a < x_i < a \\ k(-x_i - a)^m x_i < -a \end{cases}$	10	[−50, 50]	0
$F_{13}(x) = 0.1\left\{\sin(3\pi x_1)^2 + \sum_{i=1}^{n}(x_i - 1)^2\left[1 + \sin(3\pi x_i + 1)^2\right] + (x_n - 1)^2\left[1 + \sin(2\pi x_n)^2\right]\right\} + \sum_{i=1}^{n}\mu(x_i, 5, 100, 4)$	30	[−50,50]	0
$F_{14}(x) = \left(\frac{1}{500} + \sum_{j=1}^{25}\frac{1}{j + \sum_{i=1}^{2}\left(x_i - a_{ij}\right)^6}\right)^{-1}$	2	[−65, 65]	1
$F_{15}(x) = \sum_{i=1}^{11}\left[a_i - \frac{x_1\left(b_i^2 + b_i x_2\right)}{b_i^2 + b_i x_3 + x_4}\right]^2$	4	[−5, 5]	0.0003
$F_{16}(x) = 4x_1^2 - 2.1x_1^4 + \frac{1}{3}x_1^6 + x_1x_2 - 4x_2^2 + 4x_2^4$	2	[−5,5]	−1.0316

Table A3. Fixed-dimension multimodal benchmark functions [8].

Function	Dimension	Range	f_{min}
$F_{17}(x) = \left(x_2 - \frac{5.1}{4\pi^2}x_1^2 + \frac{5}{\pi}x_1 - 6\right)^2 + 10\left(1 - \frac{1}{8\pi}\right)\cos x_1 + 10$	2	[−5, 5]	0.398
$F_{18}(x) = [1 + (x_1 + x_2 + 1)^2(19 - 14x_1 + 3x_1^2 - 14x_2 + 6x_1x_2 + 3x_2^2)] \times [30 + (2x_1 - 3x_2)^2 \times (18 - 32x_1 + 12x_1^2 + 48x_2 - 36x_1x_2 + 27x_2^2)]$	2	[−2, 2]	3
$F_{19}(x) = -\sum_{i=1}^{4} c_i exp\left(-\sum_{j=1}^{3} a_{ij}\left(x_j - p_{ij}\right)^2\right)$	3	[1,3]	−3.86
$F_{20}(x) = -\sum_{i=1}^{4} c_i exp\left(-\sum_{j=1}^{6} a_{ij}\left(x_j - p_{ij}\right)^2\right)$	6	[0,1]	−3.32
$F_{21}(x) = -\sum_{i=1}^{5}\left[(X - \alpha_i)(X - \alpha_i)^T + C_i\right]^{-1}$	4	[0,10]	−10.1532
$F_{22}(x) = -\sum_{i=1}^{7}\left[(X - \alpha_i)(X - \alpha_i)^T + C_i\right]^{-1}$	4	[0,10]	−10.4028
$F_{23}(x) = -\sum_{i=1}^{1}0\left[(X - \alpha_i)(X - \alpha_i)^T + C_i\right]^{-1}$	4	[0,10]	−10.536

References

1. Wang, G.-G.; Zhao, X.; Li, K. *Metaheuristic Algorithms: Theory and Practice*; CRC Press: Boca Raton, FL, USA, 2024.
2. Munciño, D.M.; Damian-Ramírez, E.A.; Cruz-Fernández, M.; Montoya-Santiyanes, L.A.; Rodríguez-Reséndiz, J. Metaheuristic and Heuristic Algorithms-Based Identification Parameters of a Direct Current Motor. *Algorithms* **2024**, *17*, 209. [CrossRef]
3. Chen, L.; Zhao, Y.; Ma, Y.; Zhao, B.; Feng, C. Improving Wild Horse Optimizer: Integrating Multistrategy for Robust Performance across Multiple Engineering Problems and Evaluation Benchmarks. *Mathematics* **2023**, *11*, 3861. [CrossRef]
4. Ameen, A.A.; Rashid, T.A. *A Tutorial on Child Drawing Development Optimization*; Springer International Publishing AG: Muscat, Oman, 2022; pp. 1–15. Available online: http://iciitb.mcbs.edu.om/en/iciitb-home (accessed on 30 January 2023).
5. Zhou, S.; Shi, Y.; Wang, D.; Xu, X.; Xu, M.; Deng, Y. Election Optimizer Algorithm: A New Meta-Heuristic Optimization Algorithm for Solving Industrial Engineering Design Problems. *Mathematics* **2024**, *12*, 1513. [CrossRef]
6. Bibri, S.E.; Krogstie, J.; Kaboli, A.; Alahi, A. Smarter eco-cities and their leading-edge artificial intelligence of things solutions for environmental sustainability: A comprehensive systematic review. *Environ. Sci. Ecotechnol.* **2024**, *19*, 100330. [CrossRef]
7. Leiva, D.; Ramos-Tapia, B.; Crawford, B.; Soto, R.; Cisternas-Caneo, F. A Novel Approach to Combinatorial Problems: Binary Growth Optimizer Algorithm. *Biomimetics* **2024**, *9*, 283. [CrossRef]
8. Dhiman, G.; Garg, M.; Nagar, A.; Kumar, V.; Dehghani, M. A novel algorithm for global optimization: Rat swarm optimizer. *J. Ambient Intell. Humaniz. Comput.* **2021**, *12*, 8457–8482. [CrossRef]
9. Awadallah, M.A.; Al-Betar, M.A.; Braik, M.S.; Hammouri, A.I.; Doush, I.A.; Zitar, R.A. An enhanced binary Rat Swarm Optimizer based on local-best concepts of PSO and collaborative crossover operators for feature selection. *Comput. Biol. Med.* **2022**, *147*, 105675. [CrossRef]
10. Houssein, E.H.; Hosney, M.E.; Oliva, D.; Younis, E.M.G.; Ali, A.A.; Mohamed, W.M. An efficient discrete rat swarm optimizer for global optimization and feature selection in chemoinformatics. *Knowl.-Based Syst.* **2023**, *275*, 110697. [CrossRef]
11. Toolabi Moghadam, A.; Aghahadi, M.; Eslami, M.; Rashidi, S.; Arandian, B.; Nikolovski, S. Adaptive rat swarm optimization for optimum tuning of SVC and PSS in a power system. *Int. Trans. Electr. Energy Syst.* **2022**, *2022*, 4798029. [CrossRef]
12. Sayed, G.I. A novel multi-objective rat swarm optimizer-based convolutional neural networks for the diagnosis of COVID-19 disease. *Autom. Control Comput. Sci.* **2022**, *56*, 198–208. [CrossRef]
13. Zebiri, I.; Zeghida, D.; Redjimi, M. Rat swarm optimizer for data clustering. *Jordanian J. Comput. Inf. Technol.* **2022**, *8*, 1. [CrossRef]
14. Eslami, M.; Akbari, E.; Seyed Sadr, S.T; Ibrahim, B.F. A novel hybrid algorithm based on rat swarm optimization and pattern search for parameter extraction of solar photovoltaic models. *Energy Sci. Eng.* **2022**, *10*, 2689–2713. [CrossRef]
15. Manickam, P.; Girija, M.; Dutta, A.K.; Babu, P.R.; Arora, K.; Jeong, M.K.; Acharya, S. Empowering Cybersecurity Using Enhanced Rat Swarm Optimization with Deep Stack-Based Ensemble Learning Approach. *IEEE Access* **2024**, *12*, 62492–62501. [CrossRef]
16. Rahab, H.; Haouassi, H.; Souidi, M.E.H.; Bakhouche, A.; Mahdaoui, R.; Bekhouche, M. A modified binary rat swarm optimization algorithm for feature selection in Arabic sentiment analysis. *Arab. J. Sci. Eng.* **2023**, *48*, 10125–10152. [CrossRef]
17. Singla, M.K.; Gupta, J.; Alsharif, M.H.; Kim, M.-K. A modified particle swarm optimization rat search algorithm and its engineering application. *PLoS ONE* **2024**, *19*, e0296800. [CrossRef]

18. Lou, T.; Guan, G.; Yue, Z.; Wang, Y.; Tong, S. A Hybrid K-means Method based on Modified Rat Swarm Optimization Algorithm for Data Clustering. *Preprint* **2024**. [CrossRef]
19. Mzili, T.; Riffi, M.E.; Mzili, I.; Dhiman, G. A novel discrete Rat swarm optimization (DRSO) algorithm for solving the traveling salesman problem. *Decis. Mak. Appl. Manag. Eng.* **2022**, *5*, 287–299. [CrossRef]
20. Mzili, T.; Mzili, I.; Riffi, M.E. Artificial rat optimization with decision-making: A bio-inspired metaheuristic algorithm for solving the traveling salesman problem. *Decis. Mak. Appl. Manag. Eng.* **2023**, *6*, 150–176. [CrossRef]
21. Mzili, T.; Mzili, I.; Riffi, M.E. Optimizing production scheduling with the Rat Swarm search algorithm: A novel approach to the flow shop problem for enhanced decision making. *Decis. Mak. Appl. Manag. Eng.* **2023**, *6*, 16–42. [CrossRef]
22. Alruwais, N.; Alabdulkreem, E.; Khalid, M.; Negm, N.; Marzouk, R.; Al Duhayyim, M.; Balaji, P.; Ilayaraja, M.; Gupta, D. Modified rat swarm optimization with deep learning model for robust recycling object detection and classification. *Sustain. Energy Technol. Assess.* **2023**, *59*, 103397. [CrossRef]
23. Gopi, P.; Alluraiah, N.C.; Kumar, P.H.; Bajaj, M.; Blazek, V.; Prokop, L. Improving load frequency controller tuning with rat swarm optimization and porpoising feature detection for enhanced power system stability. *Sci. Rep.* **2024**, *14*, 15209. [CrossRef]
24. Ameen, A.A.; Rashid, T.A.; Askar, S. CDDO-HS: Child Drawing Development Optimization-Harmony Search Algorithm. *Appl. Sci.* **2023**, *13*, 5795. [CrossRef]
25. Ameen, A.A.; Rashid, T.A.; Askar, S. MCDDO: Overcoming Challenges and Enhancing Performance in Search Optimization. 2023. Available online: https://ouci.dntb.gov.ua/works/7qjY8BB4/ (accessed on 16 September 2024).
26. Mirjalili, S. SCA: A Sine Cosine Algorithm for solving optimization problems. *Knowl.-Based Syst.* **2016**, *96*, 120–133. [CrossRef]
27. Desuky, A.S.; Cifci, M.A.; Kausar, S.; Hussain, S.; El Bakrawy, L.M. Mud Ring Algorithm: A new meta-heuristic optimization algorithm for solving mathematical and engineering challenges. *IEEE Access* **2022**, *10*, 50448–50466. [CrossRef]
28. Houssein, E.H.; Oliva, D.; Samee, N.A.; Mahmoud, N.F.; Emam, M.M. Liver Cancer Algorithm: A novel bio-inspired optimizer. *Comput. Biol. Med.* **2023**, *165*, 107389. [CrossRef]
29. Qais, M.H.; Hasanien, H.M.; Turky, R.A.; Alghuwainem, S.; Tostado-Véliz, M.; Jurado, F. Circle search algorithm: A geometry-based metaheuristic optimization algorithm. *Mathematics* **2022**, *10*, 1626. [CrossRef]
30. Kaur, S.; Awasthi, L.K.; Sangal, A.L.; Dhiman, G. Tunicate Swarm Algorithm: A new bio-inspired based metaheuristic paradigm for global optimization. *Eng. Appl. Artif. Intell.* **2020**, *90*, 103541. [CrossRef]
31. Bairwa, A.K.; Joshi, S.; Singh, D. Dingo optimizer: A nature-inspired metaheuristic approach for engineering problems. *Math. Probl. Eng.* **2021**, *2021*, 2571863. [CrossRef]
32. Al-Betar, M.A.; Awadallah, M.A.; Braik, M.S.; Makhadmeh, S.; Doush, I.A. Elk herd optimizer: A novel nature-inspired metaheuristic algorithm. *Artif. Intell. Rev.* **2024**, *57*, 48. [CrossRef]
33. Braik, M.; Hammouri, A.; Atwan, J.; Al-Betar, M.A.; Awadallah, M.A. White Shark Optimizer: A novel bio-inspired meta-heuristic algorithm for global optimization problems. *Knowl.-Based Syst.* **2022**, *243*, 108457. [CrossRef]
34. Montgomery, D.C. *Design and Analysis of Experiments*; John Wiley & Sons: Hoboken, NJ, USA, 2017.
35. Ameen, A.A. Metaheuristic Optimazation Algorithms in Applied Science and Engineering Applications.pdf, Erbil Polytechnic University, 2024. Available online: https://epu.edu.iq/2024/03/17/metaheuristic-optimization-algorithms-in-applied-science-and-engineering-applications-2/ (accessed on 16 September 2024).
36. Hollander, M.; Wolfe, D.A.; Chicken, E. *Nonparametric Statistical Methods*; John Wiley & Sons: Hoboken, NJ, USA, 2013.
37. Iruthayarajan, M.W.; Baskar, S. Covariance matrix adaptation evolution strategy-based design of centralized PID controller. *Expert Syst. Appl.* **2010**, *37*, 5775–5781. [CrossRef]
38. Abualigah, L.; Diabat, A. Advances in sine cosine algorithm: A comprehensive survey. *Artif. Intell. Rev.* **2021**, *54*, 2567–2608. [CrossRef]
39. Eisinga, R.; Heskes, T.; Pelzer, B.; Te Grotenhuis, M. Exact p-values for pairwise comparison of Friedman rank sums, with application to comparing classifiers. *BMC Bioinform.* **2017**, *18*, 68. [CrossRef]
40. Kannan, B.K.; Kramer, S.N. An Augmented Lagrange Multiplier-Based Method for Mixed Integer Discrete Continuous Optimization and Its Applications to Mechanical Design. 1994. Available online: https://asmedigitalcollection.asme.org/mechanicaldesign/article-abstract/116/2/405/454458/An-Augmented-Lagrange-Multiplier-Based-Method-for?redirectedFrom=fulltext (accessed on 16 September 2024).
41. Çelik, Y.; Kutucu, H. Solving the Tension/Compression Spring Design Problem by an Improved Firefly Algorithm. *IDDM* **2018**, *1*, 1–7.
42. Dhiman, G. SSC: A hybrid nature-inspired meta-heuristic optimization algorithm for engineering applications. *Knowl.-Based Syst.* **2021**, *222*, 106926. [CrossRef]
43. Gandomi, A.H.; Yang, X.S.; Talatahari, S.; Alavi, A.H. Metaheuristic Algorithms in Modeling and Optimization. *Metaheuristic Appl. Struct. Infrastruct.* **2013**, *1*, 1–24. [CrossRef]
44. Fauzi, H.; Batool, U. A three-bar truss design using single-solution simulated Kalman filter optimizer. *Mekatronika J. Intell. Manuf. Mechatron.* **2019**, *1*, 98–102. [CrossRef]
45. Sandgren, E. Nonlinear integer and discrete programming in mechanical design optimization. *J. Mech. Des.* **1990**, *112*, 223–229. [CrossRef]
46. Gandomi, A.H.; Yang, X.-S.; Alavi, A.H. Cuckoo search algorithm: A metaheuristic approach to solve structural optimization problems. *Eng. Comput.* **2013**, *29*, 17–35. [CrossRef]

47. Nakajima, Y. Application of computational mechanics to tire design—Yesterday, today, and tomorrow. *Tire Sci. Technol.* **2011**, *39*, 223–244. [CrossRef]
48. Nakajima, Y.; Kadowaki, H.; Kamegawa, T.; Ueno, K. Application of a neural network for the optimization of tire design. *Tire Sci. Technol.* **1999**, *27*, 62–83. [CrossRef]
49. Ghasri, M. Benchmark Problems. MathWorks: R2022b. 2023. Available online: https://www.mathworks.com/matlabcentral/fileexchange/124810-benchmark-problems#version_history_tab (accessed on 16 September 2024).

algorithms

MDPI

Article

Comparative Analysis of Artificial Neural Networks and Evolutionary Algorithms in DEA-β-MSV Portfolio Optimization

Abdelouahed Hamdi [1,*], Arezou Karimi [2], Farshid Mehrdoust [2] and Samir Brahim Belhaouari [3]

1 Department of Mathematics and Statistics, College of Arts and Sciences, Qatar University, Doha 2713, Qatar
2 Department of Applied Mathematics, Faculty of Mathematical Sciences, University of Guilan, Rasht 41335-1914, Iran; fmehrdoust@guilan.ac.ir (F.M.)
3 Division of Information and Computing Technology, Faculty of Science and Engineering, Hamad Bin Khalifa University, Doha 5825, Qatar; sbelhaouari@hbku.edu.qa
* Correspondence: abhamdi@qu.edu.qa

Abstract: This paper proposes a hybrid methodology for portfolio optimization by integrating the data envelopment analysis (DEA) model with the mean semivariance (MSV) framework. The goal is to construct portfolios that achieve targeted returns while minimizing downside risk. The methodology comprises two stages: (1) identifying efficient stocks through DEA, where semivariance and beta (β) are employed as input risk metrics and the expected return serves as the output, and (2) determining optimal portfolio weights through the MSV model, solved using artificial neural networks (ANNs) and evolutionary algorithms. The empirical results demonstrate that portfolios optimized with ANNs exhibit significantly lower risk compared to those derived from evolutionary algorithms, highlighting the superiority of ANN-based approaches in balancing risk and return under the proposed framework. This study underscores the potential of hybrid DEA-MSV models enhanced by machine learning techniques for advanced portfolio management.

Keywords: optimal portfolio; MSV; DEA; artificial neural networks; NSGA2; SPEA2

1. Introduction

Portfolio optimization constitutes a substantial challenge within the realm of financial engineering, leading to the proliferation of various methodologies. Among these, the foundational mean variance (MV) model, as articulated by Markowitz [1], holds a position of paramount importance and lasting influence. The MV model addresses portfolio optimization by minimizing the portfolio risk subject to a target return or maximizing portfolio return subject to a defined risk tolerance. Building on this framework, Markowitz et al. [2] introduced the semivariance risk measure, arguing that its application results in the creation of portfolios that exhibit better performance relative to those constructed using variance. Subsequent research has focused on extending the mean semivariance (MSV) model to incorporate diverse real-world complexities. Huang [3] examined portfolio selection within a fuzzy environment, comparing fuzzy MV and MSV models and employing genetic algorithms for solution. Tsai and Wang [4] proposed the MSV model designed to manage uncertainty in both upside and downside risk and return. Qin et al. [5] modeled portfolio returns as fuzzy random variables, developed the MSV model with random returns, and implemented a hybrid solution algorithm. Gökgöz and Atmaca [6] optimized energy portfolios in the Turkish financial market, highlighting the influence of investor risk aversion on optimal solutions. Chen et al. [7] considered uncertain returns, transaction costs, cardinality, and boundary constraints within the MSV framework, and utilized evolutionary algorithms for solution. Hamdi et al. [8] addressed portfolio optimization under conditional value-at-risk (CVaR) by incorporating short selling, cardinality constraints, and transaction costs, and solved the resulting problem using penalty decomposition methods.

Traditional portfolio selection models primarily address unsystematic risk, exhibiting limitations in mitigating systematic risk. The beta criterion, which measures systematic risk, serves as a valuable complement to other risk measures, enhancing the overall risk management when used in combination. This is particularly relevant within the context of the capital asset pricing model (CAPM), which relates the expected return of a security to its beta risk. Researchers, including Chochola et al. [9], Chochola et al. [10], Hur and Chung [11], and Cenesizoglu and Reeves [12], have applied the CAPM model to analyze the dynamic of financial markets.

Given the vast number of assets in stock markets, the selection of efficient assets presents a significant challenge. Scientific methodologies are therefore imperative. The DEA method introduced by Charnes et al. [13] offers a non-parametric approach to assess the relative efficiency of decision-making units (DMUs) by comparing the ratio of their weighted outputs to weighted inputs. This methodology has been increasingly used in portfolio optimization. Morey and Morey [14] modeled the MV model using DEA, with variance as the input and the expected return as the output. Lamb and Tee [15] designed a skewness-MV model with DEA, evaluating mutual fund performance using variance as the input and return as the output. Branda [16] applied DEA to portfolio optimization, considering risk and return as the input and output, respectively. Xiao et al. [17] tackled the issue of uncertainty in stock returns and employed DEA for optimizing portfolios in the context of the American stock market. Zhou et al. [18] integrated DEA with multiple data sources to provide optimal portfolio support vector machines. Hamdi et al. [19] formulated a conditional risk model utilizing DEA and subsequently employed meta-heuristic algorithms for its resolution. Their research culminated in the presentation of an optimized portfolio within the Iranian stock market, demonstrating a propensity for efficient assets to receive maximal weighting within the portfolio construction.

This study aims to construct an optimal portfolio composed of efficient stocks from automotive companies while simultaneously managing the systematic risk. To this end, the MSV model is developed by integrating DEA with beta risk metrics. In contrast to traditional portfolio optimization approaches—such as mean variance or mean semivariance models that mainly depend on historical returns and risk measures like variance or semivariance—the proposed methodology assesses assets within a multidimensional analytical framework.

DEA incorporates multiple financial indicators, treating risk and financial leverage as inputs and metrics like return on equity, earnings per share, and the inverse price-to-earnings ratio as outputs. This approach provides a more comprehensive perspective on asset performance and financial health, facilitating the identification of assets that have not only generated favorable returns but are also fundamentally robust. When an asset (typically a company's stock) is deemed fundamentally robust, it signifies that the issuing company possesses sound financial characteristics and indicators that suggest long-term stability, profitability, and growth potential. These attributes pertain to the company's operational performance and intrinsic value, rather than solely its stock price in the market.

By allowing investors to prioritize criteria aligned with their objectives (e.g., earnings quality, valuation metrics, or capital structure), DEA enhances the portfolio selection process. As a result, the integration of DEA with the MSV model enables the construction of portfolios that excel not only in risk–return performance but also in foundational strength. This methodology transcends traditional optimization techniques by incorporating fundamental and operational factors, ensuring that selected assets operate efficiently and possess long-term growth potential.

Portfolio optimization problems highlight the limitations of traditional analytical techniques, which struggle to address non-linear objective functions, multiple constraints, and high-dimensional search spaces. These challenges render conventional approaches computationally inefficient or impractical in real-world scenarios. To overcome these shortcomings, the proposed methodology employs neural networks and multi-objective evolutionary algorithms (NSGA2 and SPEA2). Neural networks excel by learning intricate patterns in financial data, while evolutionary algorithms—inspired by natural selection—systematically explore vast solution spaces to identify near-optimal portfolios. This synergy enables the model to navigate the complexities of portfolio construction, prioritizing risk–return trade-offs and investor-specific criteria with greater adaptability and precision.

The remainder of this paper is organized as follows. Section 2 establishes the mathematical foundations of the DEA-β-MSV framework, detailing its components and theoretical principles. Section 3 follows with a comprehensive description of the feedforward neural network, emphasizing its architecture, training methodology, and adaptability to non-linear financial data for solving the optimization model. Section 4 then introduces the NSGA2 and SPEA2 algorithms, explaining their evolutionary mechanisms, parameter configurations, and integration into the multi-objective optimization workflow. Section 5 subsequently analyzes the empirical results derived from applying the proposed hybrid model to automotive sector portfolios, comparing the performance metrics across methodologies to evaluate their effectiveness. Finally, Section 6 synthesizes the key findings, discusses practical implications for portfolio management, and outlines potential directions for future research, reinforcing the study's contributions to adaptive and robust investment strategies.

2. Mathematical Models

This section first presents the single-index model, then examines parametric optimization methods and DEA, before formally introducing the proposed model. Some assumptions are as follows:

1. χ is a set of existing investment opportunities.
2. $\Re_i$ is the return of i-th asset on the probability space (Ω, F, P).
3. $\chi^P = \{\Re_i : i = 1, ..., n\} = \{\sum_{i=1}^{n} \Re_i \varepsilon_i : \sum_{i=1}^{n} \varepsilon_i = 1, \varepsilon_i \in \{0,1\}\}$. ε_i is the weight of each stock in the optimal portfolio.

Definition 1. *Branda [20] Functionals $\nu : L^2(\Omega) \to [0, \infty]$ are designated as general deviation measures predicated upon their satisfaction of the following axiomatic properties: translation invariance, positive homogeneity, subadditivity, and non-negativity. These criteria are formally articulated as follows:*

1. $\forall\, Y \in \chi, \forall\, C \in \mathbb{R}, \nu(Y + C) = \nu(Y)$.
2. $\nu(0) = 0$, *and* $\forall\, Y \in \chi, \forall\, \lambda > 0, \nu(\lambda Y) = \lambda \nu(Y)$.
3. $\forall\, Y_1, Y_2 \in \chi, \nu(Y_1 + Y_2) \le \nu(Y_1) + \nu(Y_2)$.
4. $\forall\, Y \in \chi, \nu(Y) \ge 0$, *with* $\nu(Y) > 0$ *for non-constant* Y.

It is pertinent to note that the axioms (2) and (3) jointly imply the convexity of such deviation measures. Prominent instantiations of this class include the standard deviation, the mean absolute deviation, and semideviations.

Definition 2. *Branda [20] Functionals $\varphi : L^2(\Omega) \to (-\infty, \infty]$ are formally designated as return measures if and only if they satisfy a specific set of axiomatic properties: translation equivariance,* *positive homogeneity, superadditivity, and monotonicity. These conditions are rigorously defined as follows:*

1. $\forall\, Y \in \chi, \forall\, C \in \mathbb{R}, \varphi(Y + C) = \varphi(Y) + C.$
2. $\varphi(0) = 0, and\, \forall\, Y \in \chi, \forall \lambda > 0, \varphi(\lambda Y) = \lambda \varphi(Y).$
3. $\forall\, Y_1, Y_2 \in \chi, \varphi(Y_1 + Y_2) \geq \varphi(Y_1) + \varphi(Y_2).$
4. $\forall\, Y_1, Y_2 \in \chi$, *if* $Y_1 \geq Y_2$ *(in the pointwise sense), then* $\varphi(Y_1) \geq \varphi(Y_2)$.

It is pertinent to observe that axioms (2) and (3) jointly imply the concavity of the functional φ*. The expectation operator is a fundamental example that demonstrably fulfills these axiomatic requirements. Furthermore, coherent risk measures, when subjected to scalar multiplication by a negative constant can be interpreted within the framework of return functionals.*

2.1. Beta Risk

Systematic risk, an intrinsic element of the total return variability observed in securities, originates from pervasive market-wide shifts and developments. This form of risk, exerting influence across a broad spectrum of securities, encompassing both equities and bonds, is an unavoidable consequence of its direct correlation with fluctuations in interest rates, market volatility, and inflationary pressures. By way of illustration, a precipitous market downturn will typically induce a corresponding depreciation in the value of a multitude of equities, whereas a rapid market ascent will conversely precipitate price appreciation. The beta coefficient (β) is conventionally employed as an index for the quantification of a security's systematic risk, and its computation proceeds as follows:

$$\beta_i = \frac{Cov(\Re_i, \Re_M)}{Var(\Re_M)}, \tag{1}$$

where $\Re_i$ represents the return of the i-th stock, and $\Re_M$ denotes the market return. Market beta is conventionally set to one. Stocks with a beta greater than one typically exhibit a higher return volatility, whereas those with a beta less than one demonstrate lower return volatility.

2.2. Linear Representation of the Single Index Model

The single index model proposes a linear relationship to represent the return of a security, expressed as follows:

$$\Re_i = A_i + \beta_i I + C_i, \tag{2}$$

where

- A_i represents the expected return of stock i;
- β_i quantifies the sensitivity of stock i to market movements (beta risk);
- C_i is a random variable with an expected value of zero and variance $\mathcal{Q}_i$;
- I represents the level of a specific index. This index can encompass various macroeconomic or market-wide factors, such as the overall stock market level, Gross National Product, a price index, or any other singular factor considered to exert the most significant influence on security returns.

2.3. Standard Optimization

Consider a set $X \subseteq \mathbb{R}^n$ within the n-dimensional space that contains all the possible decision options, represented by vectors ξ. This set of valid options, X, is usually defined by constraints related to geometry, physical laws, and real-world limitations. We use a function $f : X \to \mathbb{R}^n$ to evaluate each of these decision options, assigning a real number that indicates its desirability. For our purposes here, we will focus on finding the option that minimizes the value of f, without losing generality, since maximizing f is equivalent to minimizing f. With this in mind, the standard optimization problem can be expressed as follows:

$$\begin{aligned} \eta &= \min_{\xi} f(\xi) \\ s.t \quad & \xi \in X \subseteq \mathbb{R}^n. \end{aligned} \tag{3}$$

The goal is to identify a specific solution, represented as η, which corresponds to the minimum value of the objective function $f(\xi)$, where the possible choices for η are within the feasible set X. Typically, mathematical optimization methods are employed to locate the extreme points (either local or global) of the function f.

2.4. Quadratic Programming Representation of Single Index Model

The return on an investment portfolio (R_P) is defined as the weighted sum of the returns of its constituent securities:

$$R_P = \sum_{i=1}^{N} \varepsilon_i \Re_i,$$

where N is the number of assets and ε_i denotes the respective weighting of each asset within the portfolio. By substituting the single-index model (2) into this portfolio return equation, we obtain

$$R_P = \sum_{i=1}^{N} \varepsilon_i (A_i + \beta_i I + C_i).$$

Following Sharpe [21], this can be further simplified to

$$R_P = \sum_{i=1}^{N+1} \varepsilon_i (A_i + C_i).$$

Consequently, the expected return ($E[R_P]$) and portfolio risk ($Var[R_P]$) are determined as follows:

$$E[R_P] = \sum_{i=1}^{N+1} \varepsilon_i A_i,$$

and

$$Var[R_P] = \sum_{i=1}^{N+1} \varepsilon_i^2 Q_i.$$

Thus, the single index model naturally lends itself to a quadratic programming formulation, presented as

$$\begin{aligned} & \max \bar{\lambda} \sum_{i=1}^{N+1} \varepsilon_i A_i - \sum_{i=1}^{N+1} \varepsilon_i^2 Q_i \\ & s.t \sum_{i=1}^{N} \varepsilon_i = 1, \;\; \varepsilon_i \geq 0, \\ & \sum_{i=1}^{N} \varepsilon_i \beta_i = \varepsilon_{n+1}, \end{aligned} \tag{4}$$

where $\bar{\lambda}$ signifies the investor's level of risk aversion, and ε_{n+1} represents the weighted average responsiveness of the portfolio return (R_P) to the level of the index (I).

2.5. Parametric Optimization

Parametric optimization involves finding the solution to a standard optimization problem, but this time, the solution is expressed as a function of certain parameters. These parameters are variables that are important for making decisions but cannot be directly controlled by the designer. Within the context of systems design, examples of parameters might include design specifications that have not been finalized, environmental factors, or characteristics of other interconnected parts of the system. Conceptually, solving a parametric search problem can be understood as applying standard optimization techniques for every possible combination of the parameter value(s). Let $\Lambda \subseteq \mathbb{R}^p$ represent the set of all feasible parameter vectors, λ^*. We can extend Equation (3) to handle this parametric scenario, as follows:

$$\begin{aligned} \eta(\lambda^*) &= \min_{\xi} f(\xi, \lambda^*) \\ s.t \quad & (\xi, \lambda^*) \leq 0, \\ & h(\xi, \lambda^*) = 0, \\ & \lambda^* \in \Lambda, \\ & \xi \in \mathrm{X}. \end{aligned}$$

When dealing with parametric optimization, the objective function is optimized considering the parameter vector as an input. Specifically, $\eta(\lambda^*)$ represents the relationship between the parameter vector λ^* and the minimum value of the objective function $f(\xi, \lambda^*)$ over all feasible $\xi \in X$. Consequently, the solution is not a single point but rather a set (which could be infinite). The constraints g and h are those that include both the design variables and the parameters.

The idea of parametric optimization has been explored in fields like economics by Carlsson and Korhonen [22], Milgrom and Segal [23], often with the goal of examining how solutions change in response to variations in one or a small number of parameter variables. This kind of analysis is valuable for understanding how a system behaves when faced with uncertain parameters. Examples include changes in market demands and prices in economics, or variations in boundary conditions and system properties in engineered systems. When these kinds of fluctuating conditions are incorporated into an optimization problem, a parametric optimization approach can be employed to analyze their impact on the resulting solution.

2.6. DEA Model

Suppose there are n DMUs with input vector $x_j = (x_{1j}, \ldots, x_{mj})$ and output vector $z_j = (z_{1j}, \ldots, z_{sj})$ where $z_j \geq 0, z_j \neq 0, x_j \geq 0, x_j \neq 0$. Then, the DEA model in the nature of the input is as follows:

$$\begin{aligned} & \min \theta \\ s.t \quad & \sum_{i=1}^{n} \omega_i x_i \leq \theta x_{i0}, \qquad i = 1, \ldots m, \\ & \sum_{j=1}^{n} \omega_j z_{rj} \geq z_{r0}, \qquad r = 1, \ldots, s, \\ & \omega_j \geq 0, \qquad j = 1, \ldots, n, \end{aligned} \tag{5}$$

If $\theta = 1$, then DMU is efficient, and if $\theta < 1$, then DMU is inefficient.

In the basic DEA model, it is assumed that no specific relationship exists between inputs and outputs, and the model is non-parametric. Given our objective to extend this model for portfolio optimization, we introduce a specific assumption regarding the relationship between inputs and outputs. Since we define risk as the input and return as the output, and considering a linear relationship between these two variables, the proposed model will consequently be a parametric model.

Remark 1. *In the DEA model, the input-oriented approach evaluates the efficiency of a DMU by assuming that the unit aims to minimize its inputs while keeping outputs constant. Conversely,* *the output-oriented approach in DEA assesses the DMU efficiency based on its endeavor to maximize its outputs while holding inputs constant. In the present study, we adopted a combined approach to determine the efficiency score. Specifically, we utilized the maximum efficiency value obtained from both the input-oriented and output-oriented calculations as the final efficiency index.*

2.7. DEA Model for Risk Measurement

Consider a non-constant benchmark Y_0 belonging to the set χ. Under the condition that $\varphi_j(Y_0)$ for all $j \in \{1, \ldots, J\}$, with at least one index j satisfying this inequality, the DEA model for input-oriented risk (deviation) measurement is as follows:

$$\begin{aligned} &\min\ \theta \\ &s.t.\ \varphi_j\left(\sum_{i=1}^{n} \varepsilon_i(\Re_i)\right) \geq \varphi_j(Y_0),\ \ j = 1, \ldots, J, \\ &\nu_q\left(\sum_{i=1}^{n} \varepsilon_i(\Re_i)\right) \leq \theta.\nu_q(Y_0),\ \mathrm{q} = 1, \ldots, Q, \\ &\sum_{i=1}^{n} \varepsilon_i = 1,\ \ \varepsilon_i \geq 0,\ \ i = 1, \ldots, n, \end{aligned} \tag{6}$$

where $Y_0 \in \chi$.

2.8. Input Attitude Models with Positive or Negative Data for Risk Measures

Let the DEA model take positive or negative input values, so that the basic DEA model changes as follows:

$$\begin{aligned} &\max\theta \\ &s.t.\ \varphi_j\left(\sum_{i=1}^{n} \varepsilon_i(\Re_i)\right) \geq \varphi_j(Y_0) + \theta.\pi_j(Y_0),\ \ j = 1, \ldots, J, \\ &\nu_q\left(\sum_{i=1}^{n} \varepsilon_i(\Re_i)\right) \leq \nu_q(Y_0) - \theta.\delta_q(Y_0),\ \mathrm{q} = 1, \ldots, Q, \\ &\sum_{i=1}^{n} \varepsilon_i = 1,\ \ \varepsilon_i \geq 0,\ \ i = 1, \ldots, n, \end{aligned} \tag{7}$$

where

$$\pi_j(Y_0) = \max_{Y \in \chi}(\varphi_j(Y)) - \varphi_j(Y_0),$$

$$\delta_q(Y_0) = \nu_q(Y_0) - \min_{Y \in \chi} \nu_q(Y).$$

This model identifies the inefficiency of investment opportunity Y_0. We can write that the objective function $\min_{\theta}\frac{1-\theta}{1+\theta}$, so Y_0 is efficient if $\theta(Y_0) = 1$; otherwise, it is inefficient.

2.9. MSV Model

Semivariance, also known as downside variance, is a statistical measure of the dispersion of returns for an asset or portfolio that specifically considers deviations below a defined target value. This metric is particularly relevant for evaluating the downside risk or the potential for significant losses in investments, as it disregards positive volatility (returns above the target), unlike standard variance. This risk measure is as follows.

Let $\Re_{ti}$ represent the return of the i-th security during period t, and let k denote the investment's acceptable return threshold for the portfolio (target). Then, semivariance is defined as follows:

$$semivariance = \frac{1}{T}\sum_{t=1}^{T}\left[\left(\sum_{i=1}^{n}\Re_{ti}\varepsilon_i - k\right)^{-}\right]^2, \tag{8}$$

where

$$\left(\sum_{i=1}^{n}\Re_i\varepsilon_i - k\right)^{-} = \begin{cases}\sum_{i=1}^{n}\Re_i\varepsilon_i - k & \sum_{i=1}^{n}\Re_i\varepsilon_i < k. \\ 0 & \sum_{i=1}^{n}\Re_i\varepsilon_i \geq k.\end{cases}$$

So, the MSV model is as follows:

$$\begin{aligned} & min\frac{1}{T}\sum_{t=1}^{T}\left[\left(\sum_{i=1}^{n}\Re_{ti}\varepsilon_i - k\right)^{-}\right]^2 \\ & s.t \sum_{i=1}^{M}\varepsilon_i E(\Re_i) \geq E(\Re_p), \\ & \sum_{i=1}^{M}\varepsilon_i = 1, \\ & \varepsilon_i \geq 0, i = 1, \ldots, M, \end{aligned} \tag{9}$$

where $\Re_p$ is the portfolio return and ε_i is the weight of i-th stock in the portfolio.

The adoption of semivariance as a risk measure, particularly in financial analysis and portfolio management, stems from the understanding that investors are often more concerned with the risk of losses than with the upside volatility. Consequently, semivariance can provide a more nuanced and pertinent assessment of the risk associated with unfavorable scenarios, informing more conservative investment decision-making.

2.10. DEA Model for β-MSV

Let $DMU = (\beta, MSV)$. Then, the β-MSV model based on the DEA model is as follows:

$$\begin{aligned} & \max\ \theta \\ & s.t.\ \varphi_j\left(\sum_{i=1}^{n}\varepsilon_i(\Re_i)\right) \geq \varphi_j(Y_0) + \theta.\pi_j(Y_0),\ \ j = 1, \ldots, J, \\ & MSV_q\left(\sum_{i=1}^{n}\varepsilon_i(\Re_i)\right) \leq MSV_q(Y_0) - \theta.\delta_q(Y_0),\ \mathrm{q} = 1, \ldots, Q, \\ & \beta_q\left(\sum_{i=1}^{n}\varepsilon_i(\Re_i)\right) \leq \beta_q(Y_0) - \theta.\hat{\delta}_q(Y_0),\ \mathrm{q} = 1, \ldots, Q, \\ & \sum_{i=1}^{n}\varepsilon_i = 1,\ \ \varepsilon_i \geq 0,\ \ i = 1, \ldots, n. \end{aligned} \tag{10}$$

where

$$\pi_j(Y_0) = \max_{Y\in\chi}(\varphi_j(Y)) - \varphi_j(Y_0),$$

$$\delta_q(Y_0) = MSV_q(Y_0) - \min_{Y\in\chi}MSV_q(Y).$$

$$\hat{\delta}_q(Y_0) = \beta_q(Y_0) - \min_{Y\in\chi}\beta_q(Y).$$

Remark 2. *In accordance with Remark (1), an efficient stock is characterized in this study by its capacity to yield the minimum risk for a particular level of return, or conversely, to provide a high return for a given level of risk. This characterization similarly holds true for efficient portfolios.*

Proposition 1. *Consider the optimization problem,*

$$\min_{\omega\in\mathbb{R}^n} E[(\alpha+\omega\phi')^-]^2, \tag{11}$$

where the sentence $(\alpha+\omega\phi')^- = \min((\alpha+\omega\phi'),0)$. ϕ' *is the transpose of the matrix* $\phi = (\phi_1,\dots,\phi_n)$, w *is the asset weight vector, and* α, ϕ_i, $i = 1,\dots,n$ *are random variables by* $E[\alpha^2] < \infty$, $E[\phi_i^2] < \infty$. *If* $E[\phi_i] = 0$, *for all* i, *then the problem* (11) *accepts a solution.*

Proof. The proof of Proposition 1 can be found in Jin et al. [24]. □

Proposition 2. *Suppose the initial capital is* γ *such that* $\sum_{i=1}^{n}\varepsilon_i = \gamma$ *and* $\theta \le 1$. *Then, for any expected return of the portfolio, problem* (10) *has an optimal solution if it embraces feasible solutions.*

Proof. We can rewrite the problem (10) as follows:

$$min\frac{1}{T}\sum_{t=1}^{T}\left[\left(\sum_{i=1}^{n}\Re_{ti}\varepsilon_i - k\right)^-\right]^2$$

$$s.t.\ \varphi_j\left(\sum_{i=1}^{n}\varepsilon_i(\Re_i)\right) \ge \varphi_j(Y_0)+\theta.\pi_j(Y_0),\ \ j=1,\dots,J,$$

$$\frac{2}{1+\theta}-1\ge 0,$$

$$\beta_q\left(\sum_{i=1}^{n}\varepsilon_i(\Re_i)\right) \le \beta_q(Y_0)-\theta.\hat{\delta}_q(Y_0),\ \mathrm{q}=1,\dots,Q, \tag{12}$$

$$\sum_{i=1}^{n}\varepsilon_i = 1,\ \ \varepsilon_i \ge 0,\ \ i=1,\dots,n,$$

where

$$\pi_j(Y_0) = \max_{Y\in\chi}(\varphi_j(Y)) - \varphi_j(Y_0),$$

$$\hat{\delta}_q(Y_0) = \beta_q(Y_0) - \min_{Y\in\chi}\beta_q(Y).$$

Since the maximum value of θ occurs in $0 \le \theta \le 1$, constraint $\frac{2}{1+\theta} - 1 \ge 0$ is guaranteed. Also, the following constraint,

$$MSV_q\left(\sum_{i=1}^{n}\varepsilon_i(\Re_i)\right) \le MSV_q(Y_0) - \theta.\delta_q(Y_0),$$

in problem (10), means the lowest amount of the MSV for the highest θ. Hence, the objective function of problem (12) is guaranteed.

Now, let r_i and μ_i be the return and expected return, respectively. Then, the objective function of problem (12) can be written as follows:

$$E\left[\left(\sum_{i=1}^{n}\varepsilon_i r_i - \sum_{i=1}^{n}\varepsilon_i\mu_i\right)^-\right]^2 = E\left[\left(\sum_{i=1}^{n}\varepsilon_i(r_i-\mu_i)\right)^-\right]^2.$$

We set $\Re_i = r_i - \mu_i$. Then, based on the initial capital, we have for asset 1

$$\varepsilon_1 = \gamma - \sum_{i=2}^{n} \varepsilon_i,$$

so

$$E\left[\left(\gamma\Re_1 - \sum_{i=1}^{n} \varepsilon_i \Re_1 + \sum_{i=1}^{n} \varepsilon_i \Re_i\right)^{-}\right]^2 = E\left[\left(\gamma\Re_1 + \sum_{i=1}^{n} \varepsilon_i (\Re_i - \Re_1)\right)^{-}\right]^2.$$

Then, the objective function of problem (12) is as follows:

$$\min E\left[\left(\gamma\Re_1 + \sum_{i=1}^{n} \varepsilon_i (\Re_i - \Re_1)\right)^{-}\right]^2.$$

Let $\Re_i = \Re_1$. Then, optimal solutions are admitted by Proposition 1. □

3. Artificial Neural Network

Artificial intelligence (AI), as a broad field, aims to develop systems capable of emulating human intelligence. A pivotal approach in realizing this objective is machine learning. Machine learning involves the utilization of algorithms that enable systems to learn from data without explicit programming and autonomously enhance their performance. Neural networks (NNs), inspired by the architecture of the human brain, constitute a potent class of machine learning models and are employed in numerous advanced applications of artificial intelligence. Subsequently, this discussion will delve into a more detailed examination of a significant type of neural network, namely, feedforward neural networks. These computational structures, drawing inspiration from the operational principles of the human brain, are utilized for learning patterns inherent in data. This section will explore the mathematical underpinnings of these networks and introduce the fundamental formulas that govern their behavior across various layers (input, hidden, and output). Figure 1 shows a feedforward neural network with one hidden layer.

3.1. Feedforward Neural Network for Portfolio Semivariance Prediction

This section details the architecture of a feedforward neural network designed to predict a portfolio's semivariance.

1. Inputs and Outputs:

- **Network Input:** The input layer receives the weights of the assets in the portfolio.
- **Network Output:** The output layer, consisting of a single neuron, predicts the portfolio's semivariance, which is a continuous value.

2. Architecture and Neurons

The network features three hidden layers with 30, 20, and 10 neurons, respectively.

- **Weights (ϑ_{ji}) and Biases (b_j):** These are randomly initialized and then adjusted during the learning process using the backpropagation algorithm to minimize prediction error.
- **Activation Functions:**
  - **Hidden Layers:** We employ the hyperbolic tangent sigmoid (tansig) function, defined as $\psi(d) = \frac{1}{1+e^{-d}}$. This function compresses the output into the range $(0,1)$.
  - **Output Layer:** For this regression problem, a linear function (purelin) is used.
- **Summation Function:** The weighted sum of inputs for each neuron is automatically calculated using

$$c_j = \sum_{i=1}^{n} \vartheta_{ji} d_i + b_j.$$

3. Learning Process: Backpropagation

The learning and training process begins by feeding the network the asset weights (input) and the corresponding semivariance (output).

- **Cost Function ($J(\Theta)$):** The training function implicitly uses a cost function, typically the mean squared error (MSE), to quantify the difference between the network's predicted semivariance and the actual semivariance. The primary goal of training is to minimize this cost function.
- **Backpropagation Algorithm:** This algorithm is implemented within the training function and involves two main passes:
  - **Forward Pass:** The asset weights are propagated through the network to generate the predicted semivariance.
  - **Backward Pass:** The error at the output layer is calculated. Subsequently, the gradients of the cost function with respect to all weights and biases across all layers are computed in reverse (from the output layer towards the input layer).
- **Weight and Bias Updates:** These calculated gradients are used to update the weights and biases (e.g., $\vartheta_{ji}^{(l)} = \vartheta_{ji}^{(l)} - \kappa\frac{\partial J(\Theta)}{\partial \vartheta_{ji}^{(l)}}$) using an optimization algorithm such as Levenberg–Marquardt. The learning rate κ is an internal parameter of these optimization algorithms that governs the size of each update step.

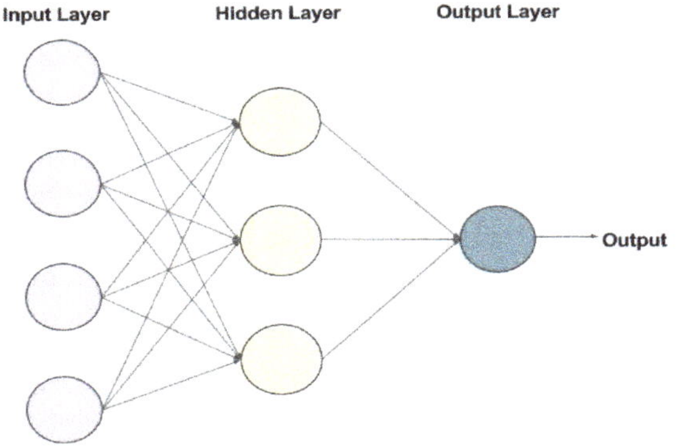

Figure 1. Feedforward neural network.

3.2. Regularization: Early Stopping

Early stopping is a crucial technique in training machine learning models, particularly neural networks. Its primary purpose is to prevent overfitting and ensure the model's better generalization capability on new, unseen data. During the training process, a model endeavors to minimize its error on the training data. However, excessive training can lead the model to "memorize" noise and specific characteristics of the training data rather than learning general patterns. This phenomenon is known as overfitting. An overfit model performs exceptionally well on data it has previously encountered but struggles with predictions on new and unknown data. Early stopping helps mitigate this issue. To implement early stopping, the original dataset is typically divided into three distinct subsets:

- **Training set:** This dataset is used to train the machine learning model, allowing it to adjust its parameters (e.g., weights in a neural network).
- **Validation set:** This portion of the data is used to monitor the model's performance during the training process. The model does not train on this data; instead, its performance is merely evaluated on it. This monitoring indicates how well the model can generalize to new data.

- **Test set:** This dataset is kept completely separate and is only used at the end of the training process for a final, unbiased evaluation of the model's performance.

The early stopping process unfolds as follows. In each training step (or "epoch"), the model's error is calculated on both the training and validation data. Initially, both errors are expected to decrease. As training progresses, the training error may continue to decline (as the model memorizes the data), but at some point, the validation error will begin to increase. This inflection point signifies the onset of overfitting. Early stopping is triggered when the validation error increases for a specified number of consecutive epochs. At this juncture, training is halted, and typically, the model's best parameters (weights and biases) from the epoch where the validation error was at its minimum are selected and saved.

4. Meta-Heuristic Algorithms

In this section, we introduce two prominent multi-objective evolutionary algorithms, NSGA2 and SPEA2, to solve the portfolio optimization problem. Both algorithms are designed to handle multiple conflicting objectives, such as maximizing the expected return while minimizing risk and transaction costs.

4.1. NSGA2 Algorithm

The genetic algorithm (GA) constitutes a robust stochastic search methodology, emulating the principles of natural selection. This algorithm integrates Darwinian principles of survival with structured random information, thereby constructing a search algorithm wherein, during each generation, the fittest individuals are selected, rather than solely the optimal ones. The GA is an iterative process designed to identify optimal solutions, manipulating a population of fixed size. The procedural steps of this algorithm are delineated as follows.

1. Creating an initial population of size N.
2. Calculation of objective function values.
3. Sorting the answers based on dominance and crowding distance.
4. Selection of parents, the act of crossing, and creating a population of children.
5. Selection of parents, mutation, and creation of the population of mutants.
6. Combining the new population with the original population.
7. Selecting members of the new main population with size N based on dominance and crowding distance.
8. If the termination conditions are not met, the second step, and, otherwise, the end.

4.2. SPEA2 Algorithm

SPEA2 uses a regular population and an archive. In this algorithm, a power value is assigned to each non-dominant answer. The power value is represented by $A(i)$. The following are defined in this algorithm.

Primary population B_0, main population in repetition B_ℓ, archive primary population $\bar{B}_0$, archive population in replication $\bar{B}_\ell$, the number of members of the main population $|B_\ell| = N$, and number of archive population members in replication $|\bar{B}_\ell| = \bar{N}$.

At first, each member i of the population $B_\ell \cup \bar{B}_\ell$ is assigned a power value with the symbol $A(i)$. $A(i)$ is the number of members of the population or archive that are defeated by i.

That is, $A(i) = |\{i | i \in B_\ell \cup \bar{B}_\ell \wedge i > j\}|, i \in B_\ell \cup \bar{B}_\ell$.

Raw Fitness

The total value of the strength of the members of the population for which i is defeated is called the raw fitness.

$$V(i) = \sum_{j \in B_\ell \cup \bar{B}_\ell, i > j} A(j).$$

$V(i)$ is an integer. In SPEA2, the more that answers are defeated by a stronger answer (the answer that overcomes more members of the population), the higher their raw fitness value will be, and in SPEA2, an answer that has a lower raw fitness is considered. That is, the less dominant answer has a higher chance of being selected. If two or more answers are the same (having the same rank but in a different distribution), then density fitting is used for the preferred answer. The fitness is calculated using the following formula:

$$F(i) = V(i) + W(i),$$

where

$$W(i) = \frac{1}{\sigma_i^u + 2}, \quad u = \sqrt{N + \bar{N}}.$$

The value of σ represents the distance of each person from their u-th nearest neighbor.

To select parents and control the archival population, if $|\bar{B}_{\ell+1}| < \bar{N}$, B_t dominant members are used to populate the population. In this method, members with better fitness are used. Unlike the classical genetic algorithm, where the new population, or the final population, is created by merging the three populations, mutation and combination, and the previous population, in SPEA2, the new population is created by the integration of the population resulting from mutation and combination, and the initial population does not affect it.

4.3. Constraint Handling in Evolutionary Optimization Problems

Section 5 details the application of the aforementioned evolutionary algorithms to solve model (12), specifically addressing the constraints $\sum_{i=1}^{n} \varepsilon_i = 1$ and $\varepsilon_i \geq 0$. Consequently, this section elucidates the methodology by which these algorithms accommodate such constraints. Both NSGA2 and SPEA2, as general-purpose multi-objective evolutionary algorithms, inherently lack specific internal mechanisms for directly handling constraints within their foundational formulations. Consequently, their application to constrained optimization problems necessitates reliance on external constraint management techniques. For the portfolio optimization problem, which is characterized by simplex feasibility constraints—namely, the non-negativity of individual weights ($\varepsilon_i \geq 0$) and the requirement that the sum of all weights equals one ($\sum_{i=1}^{n} \varepsilon_i = 1$)—we implemented a two-pronged repair-based approach, as follows:

- **Non-Negativity of Weights ($\varepsilon_i \geq 0$):** This constraint was enforced through a repair mechanism applied during the population generation process. Specifically, any portfolio weight that became negative as a result of genetic operators (crossover and mutation) was immediately set to zero. This procedure guarantees that all candidate solutions consistently adhere to the non-negativity requirement.
- **Sum of Weights Equals One ($\sum_{i=1}^{n} \varepsilon_i = 1$):** This critical constraint was meticulously managed by a dedicated normalization operator. Following the application of crossover and mutation, which can generate offspring with a sum of weights not equal to one, a normalization step was performed. In this step, the sum of all generated weights (which had already undergone the non-negativity repair) was computed, and subsequently, each individual weight was divided by this calculated sum. This iterative process ensures that the sum of weights for every portfolio precisely equals one in every generation. This repair operation is applied immediately after offspring production in each generation, thereby guaranteeing that every individual evaluated

and propagated to the subsequent generation constitutes a valid portfolio conforming to the simplex feasibility constraints.

5. Application: Iran Stock Market

This research employs an applied methodology, utilizing quantitative data within a post-event framework and drawing upon historical data from automotive enterprises. Data was collected via archival retrieval from stock exchange repositories. The statistical sample comprises data from 21 automotive firms over the period of 2017 to 2021. The selection of the Iranian automotive industry for this study was driven by several compelling factors. It represents a large and strategically important sector within Iran's economy, characterized by substantial transaction volumes and a rich availability of data, including stock prices, returns, and financial information. Furthermore, the industry's inherent high volatility and considerable susceptibility to economic policies present a challenging yet fertile ground for the rigorous testing of financial models. This dynamism provides a realistic scenario for evaluating model performance under complex market conditions. Finally, the chosen period of 2017–2021 was critical as it yielded a robust and comprehensive dataset suitable for both model training and testing, unlike preceding or subsequent periods, which contained incomplete or less pertinent data.

Statistical computations and model implementation were performed utilizing MATLAB software. Market evaluation commenced with an analysis of the automotive market index, as illustrated in Figure 2. Observations indicate a non-normal distribution of the market index. Further analysis, through the construction of a histogram, revealed a positive skew, as evidenced in Figure 2. This positive skewness suggests a potential for future appreciation in the automotive price index, implying opportunities for substantial returns for market participants. Subsequently, the statistical sample underwent analysis to ascertain the beta risk, semivariance risk, and expected returns for each stock.

Subsequently, the statistical sample underwent analysis to calculate the beta risk, semivariance risk, and expected returns for each stock from Equations (1), (8), and (13).

$$\varphi[\Re] = E[\Re] = \frac{1}{n}\sum_{i=1}^{n}\Re_i \quad (\Re_i \textit{ is the daily return}). \tag{13}$$

Additional details on the computation of daily returns and semivariance can be found in Appendix A.

The respective values for β, semivariance, and expected returns are presented in Figure 3. In Figure 3, the negative beta observed in certain stocks within automotive companies indicates a negative covariance between the returns of these stocks and the automotive market index. In other words, these stocks tend to move in the opposite direction of the automotive market. Consequently, the inclusion of such stocks in an automotive investment portfolio can mitigate the portfolio's systematic risk. Furthermore, according to Figure 3, stock #21 exhibits the lowest beta risk, while stock #1 demonstrates the lowest semivariance. Conversely, stock #8 presents the highest expected return.

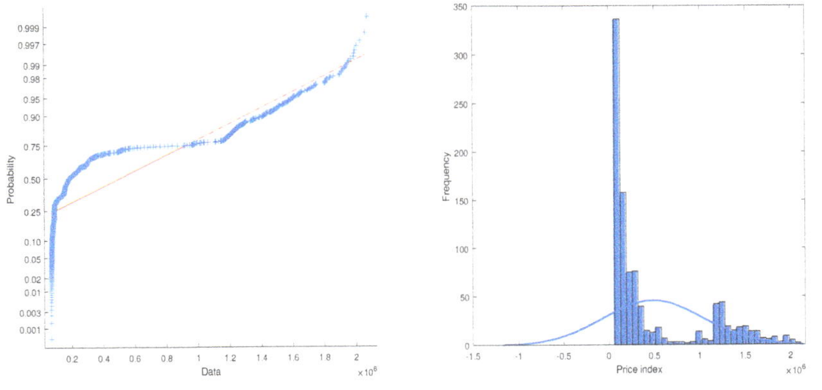

(**a**) (**b**)

Figure 2. Analyzed automotive market. (**a**) Normality test of price index. (**b**) Distribution of price index values with a long right tail.

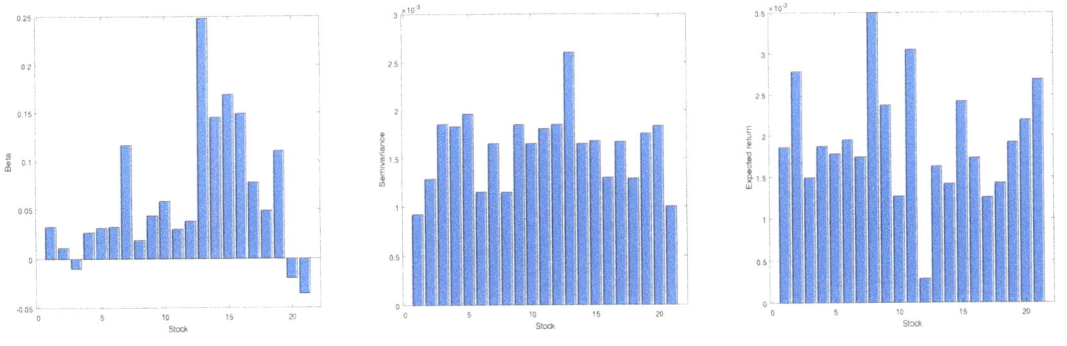

(**a**) (**b**) (**c**)

Figure 3. The risks and return of the statistical sample. (**a**) Beta of any stock. (**b**) Semivariance of any stock. (**c**) Expected return of any stock.

The proposed model (10) introduces a new parametric optimization approach, building on DEA. Its input parameters include φ (expected return), MSV (semivariance risk), and β (market risk). This model uniquely applies DEA to conceptualize risk as a consumed resource (input) and return as a generated product (output), enabling an evaluation of individual assets' relative efficiency in converting risk into return. The utilization of beta risk, in conjunction with semivariance risk, facilitates a more nuanced assessment of an asset's efficiency by incorporating systematic market risk. Assets demonstrating superior return generation for a given level of risk, or conversely, exhibiting lower risk for a given level of return, are identified as relatively more efficient and consequently receive higher efficiency scores, approaching unity. Conversely, assets exhibiting comparatively lower returns relative to their incurred risk, when juxtaposed with other assets within the dataset, are deemed inefficient and are assigned lower efficiency scores. The application of DEA efficiency scores enables the identification and subsequent exclusion of inefficient assets from a portfolio. After solving model (10), its results were presented in Table 2.

As evidenced in Table 1, with the exception of stock #8, the expected returns of the remaining stocks exhibit relative homogeneity. Consequently, as illustrated in Table 2, the DEA model identifies this particular stock as efficient due to its comparatively high return. Notably, this elevated return is achieved in conjunction with low semivariance, indicative of the stock's inherent quality. In Table 1, stock #1 demonstrates a markedly low semivariance; hence, it attains an efficiency score of 1.0 in Table 2. Furthermore, stock #21 achieves an efficiency score of 1.0 in Table 2, owing to its simultaneously low market risk and semivariance. This concurrent low level of both risk metrics for this specific stock suggests a robust underlying company performance. In environments characterized by market uncertainty, the prediction of future returns becomes increasingly complex, and risk profiles can exhibit rapid fluctuations. DEA, by evaluating the efficiency of assets in converting contemporaneous risk into contemporaneous return, offers a relatively dynamic analytical perspective. This facilitates the identification of assets that have, historically, demonstrated a superior performance relative to their accepted risk under prevailing market conditions. Moreover, the integration of beta risk alongside semivariance assists investors in considering systematic market risk in addition to downside volatility risk (semivariance). In uncertain market conditions, systematic risk can assume a more prominent role, rendering its consideration crucial for comprehensive portfolio risk management.

Furthermore, the selection of assets that exhibit efficiency in converting risk into return can indirectly lead to the inclusion of companies with stronger fundamental characteristics. Entities demonstrating greater stability and profitability are more likely to exhibit resilient performance in uncertain market conditions. Additionally, in volatile market environments, the identification and subsequent removal of assets characterized by low returns relative to their associated risk assumes heightened importance. DEA, through the provision of a quantitative efficiency score, offers investors a valuable tool for this specific purpose.

While DEA evaluates contemporaneous performance, its computation of risk and return is inherently predicated on historical data. Given the potential for past efficiency to become an unreliable predictor of future outcomes due to abrupt market shifts, it is crucial to recognize DEA as a tool for ex-post performance evaluation rather than a mechanism for forecasting. This methodology lacks the inherent capacity to directly anticipate future market fluctuations. The DEA method serves to delineate the relative efficiency of assets; however, it does not directly furnish optimal weightings for portfolio construction.

Our analysis successfully identifies efficient assets; however, allocating capital among them necessitates additional methodologies. Consequently, our subsequent goal is to determine optimal weights for these efficient stocks by solving model (12) through two different techniques: neural networks and evolutionary algorithms. The specific parameter configurations for the NSGA2 and SPEA2 algorithms can be found in Tables 3 and 4, respectively. It is crucial to distinguish that while the previous step established efficient shares considering all input and output constraints, we now simplify the problem by applying only the $\sum_{i=1}^{n} \varepsilon_i = 1, \varepsilon_i \geq 0$ constraint, thereby solving a dedicated portfolio optimization problem using these advanced methods.

Neural networks, particularly multilayer perceptrons, exhibit substantial efficacy in discerning non-linear patterns and intricate interrelationships within financial datasets. They can effectively exploit the stock efficiency metrics derived from DEA to discern underlying regularities.

Table 1. Input and output consist of the β, semivariance, and expected return.

		Inputs		Output
Asset ID	Stock Companies	β	Semivariance	Expected Return
1	CHAR1	0.0329	0.0009	0.0019
2	FNAR1	0.0113	0.0013	0.0028
3	GOST1	−0.0099	0.0019	0.0015
4	IKCO1	0.0269	0.0018	0.0019
5	KFAN1	0.0316	0.0020	0.0018
6	KHSH1	0.0323	0.0012	0.0020
7	KRIR1	0.1162	0.0017	0.0017
8	LENT1	0.0191	0.0012	0.0035
9	MESI1	0.0438	0.0018	0.0024
10	MHKM1	0.0583	0.0017	0.0013
11	MNSR1	0.0301	0.0018	0.0031
12	MSTI1	0.0383	0.0019	0.0003
13	NMOH1	0.2483	0.0026	0.0016
14	PKOD1	0.1452	0.0017	0.0014
15	RADI1	0.1694	0.0017	0.0024
16	RIIR1	0.1490	0.0013	0.0017
17	RTIR1	0.0781	0.0017	0.0013
18	SIPA1	0.0488	0.0013	0.0014
19	SZPOL1	0.1102	0.0018	0.0019
20	TMKH1	−0.0199	0.0018	0.0022
21	ZMYD1	−0.0356	0.0010	0.0027

Table 2. Efficiency of any stock.

Asset ID	Stock Companies	Efficiency
1	CHAR1	1
2	FNAR1	0.8258
3	GOST1	0.5222
4	IKCO1	0.5355
5	KFAN1	0.5096
6	KHSH1	0.8101
7	KRIR1	0.5607
8	LENT1	1
9	MESI1	0.6796
10	MHKM1	0.5606
11	MNSR1	0.8737
12	MSTI1	0.5080
13	NMOH1	0.4667
14	PKOD1	0.5607
15	RADI1	0.6928
16	RIIR1	0.7094
17	RTIR1	0.5546
18	SIPA1	0.7155
19	SZPOL 1	0.5493
20	TMKH1	0.7502
21	ZMYD1	1

We address the semivariance model for three efficient stocks using a neural network. We designed a feedforward neural network with three hidden layers. The input layer handles 1000 prospective weight vectors, while the output layer estimates portfolio semivariance. The mapping function from 1000 probable weight vectors to a portfolio's semivariance is inherently complex and non-linear. Consequently, the accurate learning of this function necessitates a network architecture beyond one or two hidden layers. This intrinsic complexity, rather than input dimensionality, determines the requisite network depth. Insufficient depth, evidenced by unsatisfactory performance, signals underfitting, for which augmenting network depth by adding more hidden layers is a suitable solution. Three hidden layers were employed, providing sufficient depth to extract abstract features from weight combinations. Neuron counts of 30, 20, and 10 were selected for these layers, respectively. This decision stemmed from suboptimal performance with fewer neurons, as lower counts reduce model capacity, hindering accurate approximations of the semivariance function. A decreasing neuron count across layers is a common and effective neural network architecture. To prevent overfitting, early stopping was implemented; training halted if the validation error worsened for six consecutive epochs without improvement.

Inherent in this design is the neural network's capacity to learn an approximation of the functional relationship between the input weights and the resultant semivariance, predicated on the provided 1000 data samples. Thus, rather than engaging in the direct computation of semivariance within the optimization process, we leverage a machine learning model to execute this task. Specifically, our aim is to utilize the neural network to approximate the objective function (semivariance) and subsequently perform a minimization procedure on this approximation to ascertain the optimal weight allocations. Fundamentally, the training of the neural network in this design entails learning a non-linear mapping function that transforms the stock weights to the corresponding semivariance value. As is visually represented in Figure 4, the attainment of a coefficient of determination (R) of 0.95 in the regression analysis conducted within the internal layers of the neural network's learning process signifies a remarkably robust model fit. This indicates that the regression model trained within these internal layers has successfully accounted for 95 percent of the variance observed in the target variable, which is likely an intermediate feature intrinsically linked to portfolio semivariance.

The neural network, in its preceding layers, has demonstrated a proficiency in extracting salient and pertinent features from the input data. These extracted features exhibit a strong correlative relationship with the target variable within the internal layers. The regression model (which may be a straightforward linear regression or a more complex non-linear formulation) trained on these extracted features has effectively modeled the relationship between these features and the target variable. This substantial goodness-of-fit within the internal layers underscores the considerable potential of the neural network to learn intricate data patterns and generate precise outputs, such as portfolio risk assessments. The declining trajectory of the mean squared error (MSE) throughout the training regimen constitutes a highly favorable and indispensable indicator of model performance. Figure 5 illustrates the model's ongoing learning and performance enhancement; with each training epoch or iteration, the neural network adaptively adjusts its internal weights to bring its output predictions (portfolio semivariance) closer to the actual or target values. A lower MSE value directly corresponds to a higher degree of accuracy in the model's predictive or generative capabilities. A consistently decreasing MSE trend signifies a continuous improvement in the model's precision. Furthermore, a stable descent in MSE suggests that the training process is converging towards an optimal (or at least a locally optimal) solution where the prediction error is minimized. The results thus obtained underscore the promising nature of employing neural networks for portfolio optimization within the capital market. The model has exhibited a notable capacity to learn complex data relationships, and its performance has demonstrably improved over the course of its training.

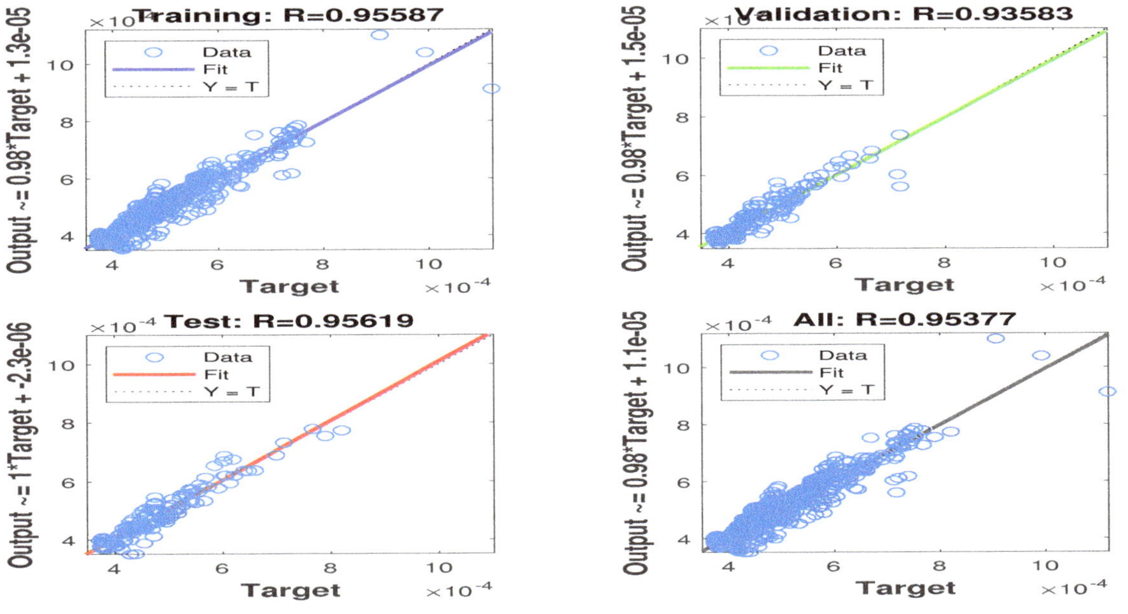

Figure 4. Feedforward neural network learning regression.

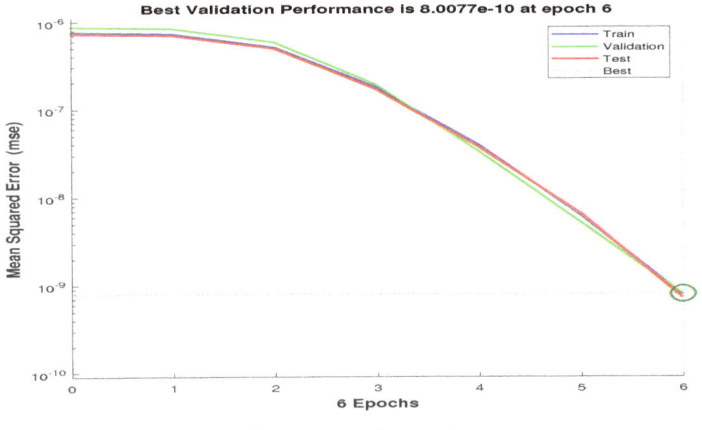

Figure 5. Performance of feedforward neural networks.

Table 3. Inputs of NSGA2.

Algorithm 1. NSGA2	Factor
Maximum Number of Iterations	100
Population Size	50
Archive Size	100
Crossover Percentage	0.7
Mutation Percentage	0.3
Mutation Rate	0.02

Table 4. Inputs of SPEA2.

Algorithm 2. SPEA2	**Factor**
Maximum Number of Iterations	100
Population Size	50
Archive Size	100
Crossover Percentage	0.7
Mutation Percentage	0.3

Upon meticulous examination of Table 5, it becomes evident that the neural network demonstrates precisely the behavior anticipated from a rational portfolio optimization process. The empirical observation that assets exhibiting the lowest semivariance (a measure of downside risk) and the highest returns ultimately receive the most substantial weight allocations within the optimized portfolio underscores the efficacy of the proposed methodology in synergistically integrating machine learning (specifically, the neural network for the explicit modeling of portfolio semivariance) with a robust optimization algorithm (in this instance, the interior-point method). In essence, the optimization algorithm, leveraging the risk predictions generated by the neural network as its objective function, has operated judiciously by preferentially allocating a greater weight to those assets that offer superior returns commensurate with lower levels of risk exposure. This principle constitutes the bedrock of a judicious investment strategy.

Furthermore, the proposed neural network has demonstrably acquired a robust understanding of the intricate relationship between the specific weight allocations assigned to individual stocks and the resultant overall risk profile of the composite portfolio. Consequently, the optimization algorithm has been effectively guided, through the informed outputs of the neural network, towards asset allocations predominantly comprising stocks characterized by desirable attributes, namely, a low risk and high return potential. This outcome is fundamentally congruent with established investment tenets that advocate for either the maximization of returns for a given, predetermined level of risk tolerance or, conversely, the minimization of risk exposure for a specified target level of return. The strategic allocation of a greater weight to assets exhibiting lower risk profiles coupled with higher return prospects represents a demonstrably rational investment heuristic that can significantly assist investors in the pursuit and attainment of their articulated financial objectives. A portfolio meticulously optimized according to these well-established principles is demonstrably more likely to exhibit superior performance over extended investment horizons when contrasted with portfolios constructed through either stochastic weight allocations or those predicated solely on intuitive judgments. This anticipated outperformance is directly attributable to the explicit focus on assets possessing both a substantial return potential and comparatively modest levels of risk. Conversely, the application of evolutionary algorithms and the traditional optimization method resulted in the predominant allocation of weight to the individual stock exhibiting the highest absolute return (stock # 8).

These alternative methodologies assigned comparatively minimal weight to stocks characterized by lower semivariance, a decision that consequently led to elevated levels of overall portfolio risk within the optimized portfolios, as explicitly detailed in Table 6. In stark contrast, the aggregate risk associated with the portfolio constructed through the informed application of the neural network is demonstrably and significantly lower. Moreover, as visually elucidated in Figures 6 and 7, the portfolios generated through the neural network approach consistently exhibit lower levels of risk exposure when directly compared to those portfolios derived from the implementation of evolutionary algorithms and the traditional method. By way of concrete illustration, for a comparable level of portfolio return approximating 0.03, the neural network-optimized portfolio exhibits a markedly lower risk level of 0.0008, whereas the risk levels associated with portfolios generated by the alternative methodologies are substantially higher, approximating 0.02.

Table 5. Weight of the efficient stock.

Asset ID	DEA-β-MSV	DEA-β-MSV-NSGA2	DEA-β-MSV-SPEA2	DEA-β-MSV-NN
1	0.0039	0.2426	0.0499	0.3687
8	0.6255	0.4400	0.6429	0.3456
21	0.3706	0.3174	0.3072	0.2857

Table 6. Return and risk of the models.

Model	Return	Risk
DEA-β- MSV-NN	0.0026	0.0008
DEA-β- MSV-NSGA2	0.0028	0.0177
DEA-β -MSV-SPEA2	0.0032	0.0203
DEA-β- MSV	0.0032	0.0242

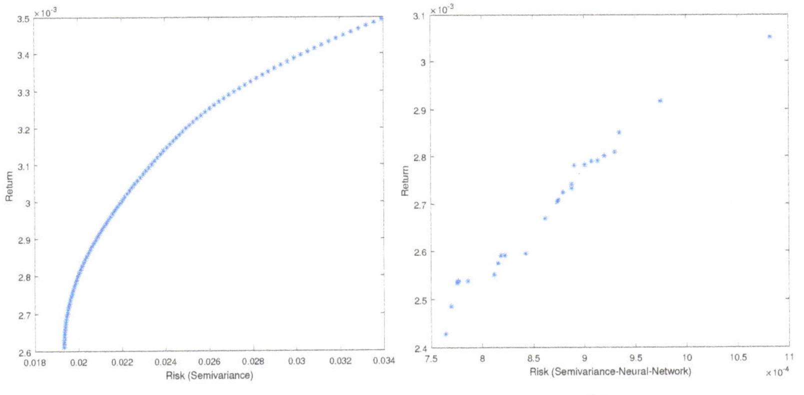

(**a**) (**b**)

Figure 6. Efficient frontier. (**a**) Frontiers of MSV model. (**b**) Frontiers of DEA-β-MSV-Neural Network model.

To further examine the portfolios generated by the introduced methods, their efficiency was evaluated using DEA in Tables 7 and 8. As observed, the neural network produced a greater number of efficient portfolios compared to the other two methods. The superior performance of the neural network in generating a larger number of efficient portfolios likely indicates its enhanced capability in learning the complex relationships between input variables (weights) and output variables (risk) within the capital market. The neural network may have identified non-linear and subtle patterns that the evolutionary algorithms were not readily able to discover, leading to the generation of portfolios that offer lower risk relative to the accepted return.

The second-place ranking of NSGA2 in terms of the number of efficient portfolios suggests that this evolutionary algorithm also demonstrated a respectable performance in identifying a set of pareto-optimal portfolios (trade-off between risk and return). NSGA2 is well-recognized for its efficacy in searching the solution space and identifying a set of non-dominated solutions (where no other solution is superior in all objectives). The last-place ranking of the SPEA2 algorithm in terms of the number of efficient portfolios may indicate that this algorithm, in comparison to the neural network and NSGA2, was less successful in identifying portfolios that simultaneously performed well across all efficiency criteria (DEA inputs and outputs). This could be attributed to the algorithm's structure, parameter settings, or its search strategy within the solution space.

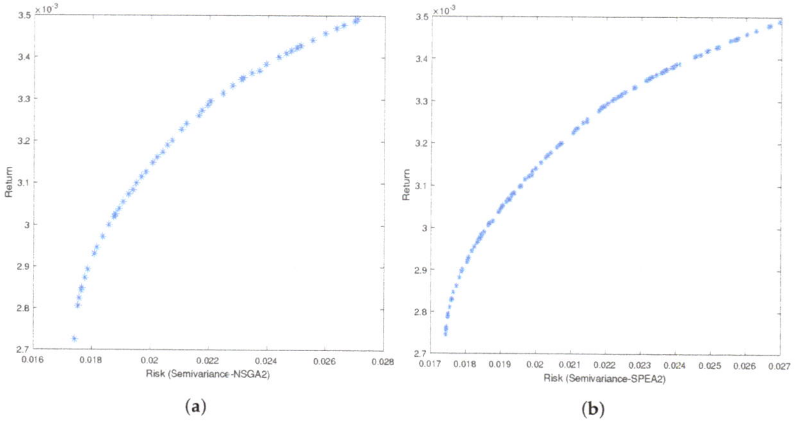

Figure 7. Efficient frontier. (**a**) Frontiers of DEA-β-MSV-NSGA2 model. (**b**) Frontiers of DEA-β-MSV-SPEA2 model.

These results can be interpreted as corroborating evidence for the significant potential of neural networks as a powerful tool in portfolio optimization. Their ability to learn intricate patterns may render them superior in identifying optimal asset allocations that establish a more favorable equilibrium between risk and return. The results demonstrate that the selection of the optimization algorithm has a significant impact on the efficiency of the final portfolios. Depending on the characteristics of the market and the investor's objectives, one method may outperform another. The performance of evolutionary algorithms such as NSGA2 and SPEA2 is highly dependent on the tuning of their parameters. It is plausible that with more precise parameter adjustments, their performance could be improved. These findings could provide impetus for exploring the combination of different optimization methods (such as utilizing the neural network's output as an initial population for a genetic algorithm) to leverage the inherent strengths of each approach.

Table 7. Input and output consist of risk and return of portfolios.

	Inputs			Outputs		
Portfolio	**NSGA2 Risk**	**SPEA2 Risk**	**NN Risk**	**NSGA2 Rerurn**	**SPEA2 Return**	**NN Return**
1	0.019	0.0215	0.0008	0.003	0.0033	0.0026
2	0.0194	0.0204	0.0004	0.0031	0.0032	0.0017
3	0.022	0.0233	0.0007	0.0033	0.0034	0.0024
4	0.0176	0.0188	0.0007	0.0028	0.003	0.0025
5	0.0188	0.0267	0.0007	0.003	0.0035	0.0024
6	0.0232	0.0241	0.0007	0.0034	0.0034	0.0025
7	0.0177	0.026	0.0007	0.002	0.0035	0.0025
8	0.0248	0.0207	0.0007	0.0034	0.0032	0.0025
9	0.025	0.0175	0.0008	0.0034	0.0028	0.0025
10	0.0188	0.0211	0.0008	0.003	0.0032	0.0025
11	0.0177	0.0203	0.0008	0.0028	0.0032	0.0025

Table 8. Efficiency of portfolios.

Portfolio	Efficiency		
	NSGA2	SPEA2	NN
1	0.9881	0.9593	1
2	1	0.9804	1
3	0.9387	0.912	0.9653
4	0.9956	0.9973	0.9814
5	0.9986	0.8193	0.9339
6	0.9171	0.8817	0.9742
7	0.6897	0.8413	1
8	0.858	0.9662	0.9922
9	0.8511	1	0.9561
10	0.9986	0.9479	0.9542
11	0.9655	0.9852	0.9725

Computational Complexity

This section provides a detailed analysis of the computational resources and time required for each component of our integrated methodology. Understanding these execution times is crucial for evaluating the practical applicability and scalability of our approach, and it clearly highlights the trade-off between solution quality and computational cost.

In portfolio optimization using a neural network, where we employed a two-stage method, the neural network training time was merely 2 s (it should be noted that early stopping was employed to prevent overfitting), and the time to find optimal weights based on the predicted semivariance was 0.5 s. Cumulatively, solving this problem with the neural network took only 2.5 s. In contrast, when solving the same problem with the NSGA2 and the SPEA2 (both configured with 100 iterations (generations) and a population size of 50), finding the optimal weights required 40 and 45 s, respectively. This direct comparison clearly demonstrates the inherent trade-off between solution quality and computational cost across the different methodologies.

The most significant advantage of our neural network-based approach emerges in its extraordinary operational speed during the second stage. After the initial neural network training (which was completed at a negligible cost of 2 s), the process of finding an optimal portfolio for a specified return is remarkably fast, typically taking only milliseconds. This unparalleled speed makes our approach highly suitable for real-time portfolio adjustments or high-frequency decision-making scenarios, where the iterative and inherently slower nature of evolutionary algorithms would be prohibitive. This highlights a clear trade-off: a very low initial training cost for the neural network, but in return, a negligible operational cost for each decision.

This characteristic provides superior speed and deterministic results compared to the stochastic nature of evolutionary algorithm executions. Among the two evolutionary algorithms tested, NSGA2 demonstrated a superior performance in terms of both solution quality and computational efficiency. NSGA2 completed its optimization on average in 40 s, indicating a slight computational advantage over SPEA2, which took approximately 45 s. Consequently, our two-stage approach, which utilizes a neural network as a semivariance surrogate model, significantly enhances the speed and scalability of portfolio optimization after initial training, making it ideal for dynamic market environments.

6. Conclusions

This study introduces a novel approach to portfolio optimization by integrating the DEA model with the MSV framework. Initially, the DEA model assesses asset efficiency using semivariance and beta as risk inputs and expected return as the output. Subsequently, an optimal portfolio is constructed from the identified efficient stocks. To determine the optimal weights of these stocks within the portfolio, the MSV model is solved using both a neural network and evolutionary algorithms (NSGA2 and SPEA2). In the neural network-based optimization method, random weights are fed into the neural network, with the portfolio's semivariance serving as the output. This output then acts as the objective function to determine the optimal asset weights. Conversely, in the evolutionary algorithm-based optimization, the MSV model is solved for the efficient stocks using the NSGA2 and SPEA2 algorithms, and their respective outcomes are compared. The findings indicate that optimizing efficient stocks with a neural network significantly reduces the portfolio risk and yields higher efficiency compared to both evolutionary algorithm-based methods. Notably, the NSGA2 algorithm outperforms SPEA2 in this context. These results suggest that neural networks are effective in discerning the inherent non-linear behavior within market data, thus offering investors a valuable tool for both forecasting and optimizing their investment portfolios. The success of the neural network in solving the semivariance model underscores its robust learning capabilities. Furthermore, the favorable performance of genetic algorithms in generating efficient portfolios opens promising avenues for future research, particularly in integrating the neural network's objective function with these algorithms for enhanced optimization and analysis.

The findings of this research are derived from data pertaining to the Iranian automotive sector over a specific period. Consequently, the unique fluctuations of the Iranian market may have influenced the model's performance. Therefore, the proposed model's performance requires further testing and validation in other markets such as developed markets with different regulatory frameworks or additional emerging markets—as well as across different asset classes, including bonds, commodities, and currencies. Additionally, generalizing these results to markets experiencing lower returns, recessionary conditions, or crises requires further validation. Hence, to further develop the DEA-β-MSV framework and broaden its applicability in more intricate market scenarios, several promising avenues for future research exist. In illiquid and noisy data markets (e.g., small-cap stocks), robust optimization can be employed to effectively manage the inherent uncertainties in return and risk estimations.

It will be crucial to integrate liquidity constraints and transaction cost models into the optimization problem, while also leveraging robust DEA models for more reliable asset filtering in such volatile conditions. For bond portfolios, the concept of downside risk can be redefined using yield spread semivariance, necessitating re-training the ANN on bond-specific data and incorporating bond-specific constraints (like duration and credit rating) into the optimization model. Furthermore, DEA inputs and outputs will require updates to accurately assess the bond efficiency. Finally, to ensure the framework's adaptability to dynamic market conditions (e.g., during crises), implementing a rolling-window strategy is vital. This involves periodic updates of the DEA, re-training or fine-tuning the ANN, and re-executing the optimization process. Moreover, regime-switching models can be utilized to dynamically adjust parameters and strategies in response to evolving market states.

Author Contributions: Formal analysis, A.H.; Methodology, A.H.; Resources, F.M.; Software, F.M. and A.K.; Supervision, S.B.B.; Validation, A.K.; Writing—original draft, A.H., F.M. and A.K.; Writing—review & editing, S.B.B. All authors have read and agreed to the published version of the manuscript.

Funding: Thanks for the funding provided by Qatar National Library (QNL).

Data Availability Statement: The data presented in this study are available on request from the corresponding author.

Conflicts of Interest: The authors declare no conflict of interest.

Appendix A. Calculation of Daily Returns and Semivariance

This appendix outlines the procedures for computing daily asset returns and the semivariance used in portfolio optimization under the DEA-β-MSV framework.

Appendix A.1. Daily Returns Calculation

Daily returns for each asset were calculated using log-returns (logarithmic returns). Log-returns are preferred over simple returns in financial time series analysis, particularly when aggregating returns over multiple periods, as they are additive across time and approximate continuous compounding. The daily log-return for an asset i on day t, denoted as $\Re_{it}$, is computed as

$$\Re_{it} = \ln \frac{P_{i,t}}{P_{i,t-1}},$$

where

- $P_{i,t}$ is the closing price of asset i on day t;
- $P_{i,t-1}$ is the closing price of asset i on day $t-1$;
- ln denotes the natural logarithm.

Appendix A.2. Semivariance Computation

Semivariance is a measure of the downside risk, focusing only on the deviations of returns that fall below a specified benchmark or target. It quantifies the volatility associated with negative returns, providing a more focused perspective on potential losses compared to the total variance. The semivariance for a portfolio (or individual asset) is calculated based on historical daily returns that fall below a predefined threshold, typically, the mean return or a target return. In this study, semivariance was computed relative to the mean daily return of the asset/portfolio.

The semivariance of asset i, denoted as SV_i, is computed over T daily observations, as follows:

$$\text{SV}_i = \frac{1}{T} \sum_{t=1}^{T} (\min(0, \Re_{i,t} - \bar{R}_i))^2,$$

where

- T is the total number of daily observations.
- $\Re_{i,t}$ is the daily return of asset i on day t.
- $\bar{R}_i$ is the mean daily return of asset i over the period T, calculated as $\bar{R}_i = \frac{1}{T} \sum_{t=1}^{T} \Re_{i,t}$.
- The $\min(0, \cdot)$ function ensures that only negative deviations from the mean are considered. Deviations above the mean are treated as zero.

For a portfolio P with N assets and weights ε_j, the portfolio daily return $R_{P,t} = \sum_{j=1}^{N} \varepsilon_j \Re_{j,t}$. The semivariance for the portfolio, SV_P, is then calculated similarly:

$$\text{SV}_P = \frac{1}{T} \sum_{t=1}^{T} (\min(0, R_{P,t} - \bar{R}_P))^2,$$

where $\bar{R}_P$ is the mean daily return of the portfolio over the period T.

References

1. Markowitz, H.M. *Portfolio Selection: Efficient Diversification of Investments;* John Wiley: Hoboken, NJ, USA, 1959.
2. Markowitz, H.; Todd, P.; Xu, G.; Yamane, Y. Computation of mean-semivariance efficient sets by the critical line algorithm. *Ann. Oper. Res.* **1993**, *45*, 307–317. [CrossRef]
3. Huang, X. Mean-semivariance models for fuzzy portfolio selection. *J. Comput. Appl. Math.* **2008**, *217*, 1–8. [CrossRef]
4. Tsai, M.F.; Wang, C.J. Post-modern portfolio theory for information retrieval. *Procedia Comput. Sci.* **2012**, *13*, 80–85. [CrossRef]

5. Qin, Z.; Wang, D.Z.; Li, X. Mean-semivariance models for portfolio optimization problem with mixed uncertainty of fuzziness and randomness. *Int. J. Uncertain. Fuzziness Knowl.-Based Syst.* **2013**, *21* (Suppl. S1), 127–139. [CrossRef]
6. Gökgöz, F.; Atmaca, M.E. Portfolio optimization under lower partial moments in emerging electricity markets: Evidence from Turkey. *Renew. Sustain. Energy Rev.* **2017**, *67*, 437–449. [CrossRef]
7. Chen, W.; Li, D.; Lu, S.; Liu, W. Multi-period mean–semivariance portfolio optimization based on uncertain measure. *Soft Comput.* **2019**, *23*, 6231–6247. [CrossRef]
8. Hamdi, A.; Khodamoradi, T.; Salahi, M. A penalty decomposition algorithm for the extended mean–variance–CVaR portfolio optimization problem. *Discret. Math. Algorithms Appl.* **2023**, *16*, 2350021. [CrossRef]
9. Chochola, O.; Hušková, M.; Prášková, Z.; Steinebach, J.G. Robust monitoring of CAPM portfolio betas. *J. Multivar. Anal.* **2013**, *115*, 374–395. [CrossRef]
10. Chochola, O.; Hušková, M.; Prášková, Z.; Steinebach, J.G. Robust monitoring of CAPM portfolio betas II. *J. Multivar. Anal.* **2014**, *132*, 58–81. [CrossRef]
11. Hur, S.K.; Chung, C.Y. Revisiting CAPM betas in an incomplete market: Evidence from the Korean stock market. *Financ. Res. Lett.* **2017**, *21*, 241–248. [CrossRef]
12. Cenesizoglu, T.; Reeves, J.J. CAPM, components of beta and the cross section of expected returns. *J. Empir. Financ.* **2018**, *49*, 223–246. [CrossRef]
13. Charnes, A.; Cooper, W.W.; Rhodes, E. Measuring the efficiency of decision making units. *Eur. J. Oper. Res.* **1978**, *2*, 429–444. [CrossRef]
14. Morey, M.R.; Morey, R.C. Mutual fund performance appraisals: A multi-horizon perspective with endogenous benchmarking. *Omega* **1999**, *27*, 241–258. [CrossRef]
15. Lamb, J.D.; Tee, K.H. Data envelopment analysis models of investment funds. *Eur. J. Oper. Res.* **2012**, *216*, 687–696. [CrossRef]
16. Branda, M. Diversification-consistent data envelopment analysis with general deviation measures. *Eur. J. Oper. Res.* **2013**, *226*, 626–635. [CrossRef]
17. Xiao, H.; Ren, T.; Zhou, Z.; Liu, W. Parameter uncertainty in estimation of portfolio efficiency: Evidence from an interval diversification-consistent DEA approach. *Omega* **2021**, *103*, 102357. [CrossRef]
18. Zhou, Z.; Gao, M.; Xiao, H.; Wang, R.; Liu, W. Big data and portfolio optimization: A novel approach integrating DEA with multiple data sources. *Omega* **2021**, *104*, 102479. [CrossRef]
19. Hamdi, A.; Karimi, A.; Mehrdoust, F.; Belhaouari, S.B. Portfolio Selection Problem Using CVaR Risk Measures Equipped with DEA, PSO, and ICA Algorithms. *Mathematics* **2022**, *10*, 2808. [CrossRef]
20. Branda, M. Reformulations of input–output oriented DEA tests with diversification. *Oper. Res. Lett.* **2013**, *41*, 516–520. [CrossRef]
21. Sharpe, W.F. A simplified model for portfolio analysis. *Manag. Sci.* **1963**, *9*, 277–293. [CrossRef]
22. Carlsson, C.; Korhonen, P. A parametric approach to fuzzy linear programming. *Fuzzy Sets Syst.* **1986**, *20*, 17–30. [CrossRef]
23. Milgrom, P.; Segal, I. Envelope theorems for arbitrary choice sets. *Econometrica* **2002**, *70*, 583-601. [CrossRef]
24. Jin, H.; Markowitz, H.; Zhou, X.Y. A note on semivariance. Mathematical Finance. *Int. J. Math. Stat. Financ. Econ.* **2006**, *16*, 53–61.

MDPI AG
Grosspeteranlage 5
4052 Basel
Switzerland
Tel.: +41 61 683 77 34

Algorithms Editorial Office
E-mail: algorithms@mdpi.com
www.mdpi.com/journal/algorithms

www.ingramcontent.com/pod-product-compliance
Lightning Source LLC
Chambersburg PA
CBHW061559210926
51850CB00009B/2154
9783725864164